MASTERMIND

MARIA KONNIKOVA

MASTERMIND

HOW TO THINK LIKE
SHERLOCK HOLMES

VIKING

VIKING
Published by the Penguin Group
Penguin Group (USA) Inc., 375 Hudson Street,
New York, New York 10014, U.S.A.
Penguin Group (Canada), 90 Eglinton Avenue East, Suite 700, Toronto, Ontario,
Canada M4P 2Y3 (a division of Pearson Penguin Canada Inc.)
Penguin Books Ltd, 80 Strand, London WC2R 0RL, England
Penguin Ireland, 25 St. Stephen's Green, Dublin 2, Ireland
(a division of Penguin Books Ltd)
Penguin Group (Australia), 707 Collins Street, Melbourne, Victoria 3008
Australia (a division of Pearson Australia Group Pty Ltd)
Penguin Books India Pvt Ltd, 11 Community Centre,
Panchsheel Park, New Delhi–110 017, India
Penguin Group (NZ), 67 Apollo Drive, Rosedale, Auckland 0632,
New Zealand (a division of Pearson New Zealand Ltd)
Penguin Books, Rosebank Office Park, 181 Jan Smuts Avenue,
Parktown North 2193, South Africa
Penguin China, B7 Jaiming Center, 27 East Third Ring Road North,
Chaoyang District, Beijing 100020, China

Penguin Books Ltd, Registered Offices: 80 Strand, London WC2R 0RL, England

First published in 2013 by Viking Penguin, a member of Penguin Group (USA) Inc.

3 5 7 9 10 8 6 4

Portions of this book appeared in a different form on the website
Big Think (www.bigthink.com) and in *Scientific American*.

Photograph credits:
Page 43 (bottom left): United States Government
43 (bottom right): Wikimichels (Creative Commons Attribution-Share Alike 3.0)
234 (bottom left): Biophilia curiosus (Creative Commons Attribution 3.0)
234 (bottom right): Brandon Motz (Creative Commons Attribution 2.0)

LIBRARY OF CONGRESS CATALOGING IN PUBLICATION DATA
Konnikova, Maria.
Mastermind : how to think like Sherlock Holmes / Maria Konnikova.
p. cm.
Includes index.
ISBN 978-0-670-02657-9 (hardback)
1. Logic. 2. Reasoning. I. Title.
BC108.K58 2013
153.4—dc23
2012035455

Printed in the United States of America
Set in Minion Pro   Designed by Francesca Belanger

To Geoff

Choice of attention—to pay attention to this and ignore that—is to the inner life what choice of action is to the outer. In both cases man is responsible for his choice and must accept the consequences. As Ortega y Gasset said: "Tell me to what you pay attention, and I will tell you who you are."

—W. H. AUDEN

CONTENTS

MASTERMIND

Prelude

When I was little, my dad used to read us Sherlock Holmes stories before bed. While my brother often took the opportunity to fall promptly asleep on his corner of the couch, the rest of us listened intently. I remember the big leather armchair where my dad sat, holding the book out in front of him with one arm, the dancing flames from the fireplace reflecting in his black-framed glasses. I remember the rise and fall of his voice as the suspense mounted beyond all breaking points, and finally, finally, at long last the awaited solution, when it all made sense and I'd shake my head, just like Dr. Watson, and think, *Of course; it's all so simple now that he says it.* I remember the smell of the pipe that my dad himself would smoke every so often, a fruity, earthy mix that made its way into the folds of the leather chair, and the outlines of the night through the curtained French windows. His pipe, of course, was ever-so-slightly curved just like Holmes's. And I remember that final slam of the book, the thick pages coming together between the crimson covers, when he'd announce, "That's it for tonight." And off we'd go—no matter how much begging and pleading we'd try and what sad faces we'd make—upstairs, up to bed.

And then there's the one thing that wedged its way so deeply into my brain that it remained there, taunting me, for years to come, when the rest of the stories had long since faded into some indeterminate background and the adventures of Holmes and his faithful Boswell were all but forgotten: the steps.

The steps to 221B Baker Street. How many were there? It's the question Holmes brought before Watson in "A Scandal in Bohemia," and a question that never once since left my mind. As Holmes and Watson sit in their matching armchairs, the detective instructs the doctor on the difference between seeing and observing. Watson is baffled. And then, all at once everything becomes crystal clear.

> "When I hear you give your reasons," [Watson] remarked, "the thing always appears to me to be so ridiculously simple that I could easily do it myself, though at each successive instance of your reasoning, I am baffled until you explain your process. And yet I believe that my eyes are as good as yours."
>
> "Quite so," [Holmes] answered, lighting a cigarette, and throwing himself down into an armchair. "You see, but you do not observe. The distinction is clear. For example, you have frequently seen the steps which lead up from the hall to this room."
>
> "Frequently."
>
> "How often?"
>
> "Well, some hundreds of times."
>
> "Then how many are there?"
>
> "How many? I don't know."
>
> "Quite so! You have not observed. And yet you have seen. That is just my point. Now, I know that there are seventeen steps, because I have both seen and observed."

When I first heard it, on one firelit, pipe-smoke-filled evening, the exchange shook me. Feverishly, I tried to remember how many steps there were in our own house (I had not the faintest idea), how many led up to our front door (I drew a beautiful blank), how many led down to the basement (ten? twenty? I couldn't even approximate). And for a long time afterward, I tried to count stairs and steps whenever I could, lodging the proper number in my memory in case anyone ever called upon me to report. I'd make Holmes proud.

Of course, I'd promptly forget each number I so diligently tried to remember—and it wasn't until later that I realized that by focusing so intently on memorization, I'd missed the point entirely. My efforts had been doomed from the start.

What I couldn't understand then was that Holmes had quite a bit more than a leg up on me. For most of his life, he had been honing a method of mindful interaction with the world. The Baker Street steps?

Just a way of showing off a skill that now came so naturally to him that it didn't require the least bit of thought. A by-the-way manifestation of a process that was habitually, almost subconsciously, unfolding in his constantly active mind. A trick, if you will, of no real consequence, and yet with the most profound implications if you stopped to consider what made it possible. A trick that inspired me to write an entire book in its honor.

The idea of mindfulness itself is by no means a new one. As early as the end of the nineteenth century, William James, the father of modern psychology, wrote that "the faculty of voluntarily bringing back a wandering attention, over and over again, is the very root of judgment, character, and will. . . . An education which should improve this faculty would be the education par excellence." That faculty, at its core, is the very essence of mindfulness. And the education that James proposes, an education in a mindful approach to life and to thought.

In the 1970s, Ellen Langer demonstrated that mindfulness could reach even further than improving "judgment, character, and will." A mindful approach could go as far as to make elderly adults feel and act younger—and could even improve their vital signs, such as blood pressure, and their cognitive function. In recent years, studies have shown that meditation-like thought (an exercise in the very attentional control that forms the center of mindfulness), for as little as fifteen minutes a day, can shift frontal brain activity toward a pattern that has been associated with more positive and more approach-oriented emotional states, and that looking at scenes of nature, for even a short while, can help us become more insightful, more creative, and more productive. We also know, more definitively than we ever have, that our brains are not built for multitasking—something that precludes mindfulness altogether. When we are forced to do multiple things at once, not only do we perform worse on all of them but our memory decreases and our general well-being suffers a palpable hit.

But for Sherlock Holmes, mindful presence is just a first step. It's a means to a far larger, far more practical and practically gratifying goal. Holmes provides precisely what William James had prescribed: an education in improving our faculty of mindful thought and in using it in order to accomplish more, think better, and decide more optimally. In its broadest application, it is a means for improving overall decision making and judgment ability, starting from the most basic building block of your own mind.

What Homes is really telling Watson when he contrasts seeing and observing is to never mistake mindlessness for mindfulness, a passive approach with an active involvement. We see automatically: a stream of sensory inputs that requires no effort on our part, save that of opening our eyes. And we see unthinkingly, absorbing countless elements from the world without necessarily processing what those elements might be. We may not even realize we've seen something that was right before our eyes. But when we observe, we are forced to pay attention. We have to move from passive absorption to active awareness. We have to engage. It's true for everything—not just sight, but each sense, each input, each thought.

All too often, when it comes to our own minds, we are surprisingly mindless. We sail on, blithely unaware of how much we are missing, of how little we grasp of our own thought process—and how much better we could be if only we'd taken the time to understand and to reflect. Like Watson, we plod along the same staircase tens, hundreds, thousands of times, multiple times a day, and we can't begin to recall the most mundane of details about them (I wouldn't be surprised if Holmes had asked about color instead of number of steps and had found Watson equally ignorant).

But it's not that we aren't capable of doing it; it's just that we don't choose to do it. Think back to your childhood. Chances are, if I asked you to tell me about the street where you grew up, you'd be able to recall any number of details. The colors of the houses. The quirks of the neighbors. The smells of the seasons. How different the street was at different times of day. Where you played. Where you walked. Where you were afraid of walking. I bet you could go on for hours.

As children, we are remarkably aware. We absorb and process information at a speed that we'll never again come close to achieving. New sights, new sounds, new smells, new people, new emotions, new experiences: we are learning about our world and its possibilities. Everything is new, everything is exciting, everything engenders curiosity. And because of the inherent newness of our surroundings, we are exquisitely alert; we are absorbed; we take it all in. And what's more, we remember: because we are motivated and engaged (two qualities we'll return to repeatedly), we not only take the world in more fully than we are ever likely to do again, but we store it for the future. Who knows when it might come in handy?

But as we grow older, the blasé factor increases exponentially. Been there, done that, don't need to pay attention to this, and when in the world will I ever need to know or use that? Before we know it, we have shed that innate attentiveness, engagement, and curiosity for a host of passive, mindless habits. And even when we want to engage, we no longer have that childhood luxury. Gone are the days where our main job was to learn, to absorb, to interact; we now have other, more pressing (or so we think) responsibilities to attend to and demands on our minds to address. And as the demands on our attention increase—an all too real concern as the pressures of multitasking grow in the increasingly 24/7 digital age—so, too, does our *actual* attention decrease. As it does so, we become less and less able to know or notice our own thought habits, and more and more allow our minds to dictate our judgments and decisions, instead of the other way around. And while that's not inherently a bad thing—in fact, we'll be talking repeatedly about the need to automate certain processes that are at first difficult and cognitively costly—it is dangerously close to mindlessness. It's a fine line between efficiency and thoughtlessness—and one that we need to take care not to cross.

You've likely had the experience where you need to deviate from a stable routine only to find that you've somehow forgotten to do so. Let's say you need to stop by the drugstore on your way home. All day long, you remember your errand. You rehearse it; you even picture the extra turn you'll have to take to get there, just a quick step from your usual route. And yet somehow, you find yourself back at your front door, without having ever stopped off. You've forgotten to take that turn and you don't even remember passing it. It's the habit mindlessly taking over, the routine asserting itself against whatever part of your mind knew that it needed to do something else.

It happens all the time. You get so set in a specific pattern that you go through entire chunks of your day in a mindless daze (and if you are still thinking about work? worrying about an email? planning ahead for dinner? forget it). And that automatic forgetfulness, that ascendancy of routine and the ease with which a thought can be distracted, is just the smallest part—albeit a particularly noticeable one, because we have the luxury of realizing that we've forgotten to do something—of a much larger phenomenon. It happens much more regularly than we can point to—and more often than not, we aren't even aware of our own mindlessness. How many thoughts float in and out of your head without your stopping to identify them? How many ideas and insights have escaped because you forgot to pay attention? How many decisions or judgments have you made without realizing how or why you made them, driven by some internal default settings of whose existence you're only vaguely, if at all, aware? How many days have gone by where you suddenly wonder what exactly you did and how you got to where you are?

This book aims to help. It takes Holmes's methodology to explore and explain the steps necessary for building up habits of thought that will allow you to engage mindfully with yourself and your world as a matter of course. So that you, too, can offhandedly mention that number of steps to dazzle a less-with-it companion.

So, light that fire, curl up on that couch, and prepare once more to join Sherlock Holmes and Dr. John H. Watson on their adventures through the crime-filled streets of London—and into the deepest crevices of the human mind.

PART ONE

UNDERSTANDING (YOURSELF)

CHAPTER ONE

The Scientific Method
of the Mind

Something sinister was happening to the farm animals of Great Wyrley. Sheep, cows, horses—one by one, they were falling dead in the middle of the night. The cause of death: a long, shallow cut to the stomach that caused a slow and painful bleeding. Farmers were outraged; the community, shocked. Who would want to cause such pain to defenseless creatures?

The police thought they had their answer: George Edalji, the half-Indian son of the local vicar. In 1903, twenty-seven-year-old Edalji was sentenced to seven years of hard labor for one of the sixteen mutilations, that of a pony whose body had been found in a pit near the vicar's residence. Little did it matter that the vicar swore his son was asleep at the time of the crime. Or that the killings continued after George's imprisonment. Or, indeed, that the evidence was largely based on anonymous letters that George was said to have written—in which he implicated himself as the killer. The police, led by Staffordshire chief constable captain George Anson, were certain they had their man.

Three years later, Edalji was released. Two petitions protesting his innocence—one, signed by ten thousand people, the other, from a group of three hundred lawyers—had been sent to the Home Office, citing a lack of evidence in the case. And yet, the story was far from over. Edalji may have been free in person, but in name, he was still guilty. Prior to his arrest he had been a solicitor. Now he could not be readmitted to his practice.

In 1906, George Edalji caught a lucky break: Arthur Conan Doyle, the famed creator of Sherlock Holmes, had become interested in the case. That winter, Conan Doyle agreed to meet Edalji at the Grand Hotel, at Charing Cross. And there, across the lobby, any lingering doubts Sir

Arthur may have had about the young man's innocence were dispelled. As he later wrote:

> He had come to my hotel by appointment, but I had been delayed, and he was passing the time by reading the paper. I recognized my man by his dark face, so I stood and observed him. He held the paper close to his eyes and rather sideways, proving not only a high degree of myopia, but marked astigmatism. The idea of such a man scouring fields at night and assaulting cattle while avoiding the watching police was ludicrous. . . . There, in a single physical defect, lay the moral certainty of his innocence.

But though Conan Doyle himself was convinced, he knew it would take more to capture the attention of the Home Office. And so, he traveled to Great Wyrley to gather evidence in the case. He interviewed locals. He investigated the scenes of the crimes, the evidence, the circumstances. He met with the increasingly hostile Captain Anson. He visited George's old school. He reviewed old records of anonymous letters and pranks against the family. He traced the handwriting expert who had proclaimed that Edalji's hand matched that of the anonymous missives. And then he put his findings together for the Home Office.

The bloody razors? Nothing but old rust—and, in any case, incapable of making the type of wounds that had been suffered by the animals. The dirt on Edalji's clothes? Not the same as the dirt in the field where the pony was discovered. The handwriting expert? He had previously made mistaken identifications, which had led to false convictions. And, of course, there was the question of the eyesight: could someone with such astigmatism and severe myopia really navigate nocturnal fields in order to maim animals?

In the spring of 1907, Edalji was finally cleared of the charge of animal slaughter. It was less than the complete victory for which Conan Doyle had hoped—George was not entitled to any compensation for his arrest and jail time—but it was something. Edalji was readmitted to his legal practice. The Committee of Inquiry found, as summarized by Conan Doyle, that "the police commenced and carried on their investigations, not for the purpose of finding out who was the guilty party, but for the purpose of finding evidence against Edalji, who they were already sure was the guilty man." And in August of that year, England saw the creation of its first court of appeals, to deal with future miscarriages of justice in a more systematic fashion. The Edalji case was widely considered one of the main impetuses behind its creation.

Conan Doyle's friends were impressed. None, however, hit the nail on the head quite so much as the novelist George Meredith. "I shall not mention the name which must have become wearisome to your ears," Meredith told Conan Doyle, "but the creator of the marvellous Amateur Detective has shown what he can do in the life of breath." Sherlock Holmes might have been fiction, but his rigorous approach to thought was very real indeed. If properly applied, his methods could leap off the page and result in tangible, positive changes—and they could, too, go far beyond the world of crime.

Say the name Sherlock Holmes, and doubtless, any number of images will come to mind. The pipe. The deerstalker. The cloak. The violin. The hawklike profile. Perhaps William Gillette or Basil Rathbone or Jeremy Brett or any number of the luminaries who have, over the years, taken up Holmes's mantle, including the current portrayals by Benedict Cumberbatch and Robert Downey, Jr. Whatever the pictures your mind brings up, I would venture to guess that the word *psychologist* isn't one of them. And yet, perhaps it's time that it was.

Holmes was a detective second to none, it is true. But his insights into the human mind rival his greatest feats of criminal justice. What Sherlock Holmes offers isn't just a way of solving crime. It is an entire way of thinking, a mindset that can be applied to countless enterprises far removed from the foggy streets of the London underworld. It is an approach born out of the scientific method that transcends science and crime both and can serve as a model for thinking, a way of being, even, just as powerful in our time as it was in Conan Doyle's. And that, I would argue, is the secret to Holmes's enduring, overwhelming, and ubiquitous appeal.

When Conan Doyle created Sherlock Holmes, he didn't think much of his hero. It's doubtful that he set out intentionally to create a model for thought, for decision making, for how to structure, lay out, and solve problems in our minds. And yet that is precisely what he did. He created, in effect, the perfect spokesperson for the revolution in science and thought that had been unfolding in the preceding decades and would continue into the dawn of the new century. In 1887, Holmes became a new kind of detective, an unprecedented thinker who deployed his mind in unprecedented ways. Today, Holmes serves an ideal model for how we can think better than we do as a matter of course.

In many ways, Sherlock Holmes was a visionary. His explanations, his methodology, his entire approach to thought presaged developments in psychology and neuroscience that occurred over a hundred years after his birth—and over eighty years after his creator's death. But somehow, too, his way of thought seems almost inevitable, a clear product of its time and place in history. If the scientific method was coming into its prime in all manner of thinkings and doings—from evolution to radiography, general relativity to the discovery of germs and anesthesia, behaviorism to psychoanalysis—then why ever not in the principles of thought itself?

In Arthur Conan Doyle's own estimation, Sherlock Holmes was meant from the onset to be an embodiment of the scientific, an ideal that we could aspire to, if never emulate altogether (after all, what are ideals for if not to be just a little bit out of reach?). Holmes's very name speaks at once of an intent beyond a simple detective of the old-fashioned sort: it is very likely that Conan Doyle chose it as a deliberate tribute to one of his childhood idols, the philosopher-doctor Oliver Wendell Holmes, Sr., a figure known as much for his writing as for his contributions to medical practice. The detective's character, in turn, was modeled after another mentor, Dr. Joseph Bell, a surgeon known for his powers of close observation. It was said that Dr. Bell could tell from a single glance that a patient was a recently discharged noncommissioned officer in a Highland regiment, who had just returned from service in Barbados, and that he tested routinely his students' own powers of perception with methods that included self-experimentation with various noxious substances. To students of Holmes, that may all sound rather familiar. As Conan Doyle wrote to Bell, "Round the centre of deduction and inference and observation which I have heard you inculcate, I have tried to build up a man who pushed the thing as far as it would go—further occasionally. . . ." It is here, in observation and inference and deduction, that we come to the heart of what it is exactly that makes Holmes who he is, distinct from every other detective who appeared before, or indeed, after: the detective who elevated the art of detection to a precise science.

We first learn of the quintessential Sherlock Holmes approach in *A Study in Scarlet*, the detective's first appearance in the public eye. To Holmes, we soon discover, each case is not just a case as it would appear to the officials of Scotland Yard—a crime, some facts, some persons of interest, all coming together to bring a criminal to justice—but is something both more and less. More, in that it takes on a larger, more general significance, as an object of broad speculation and inquiry, a scientific conundrum, if you will. It has contours that inevitably were seen before in earlier problems and will certainly repeat again, broader principles that can apply to other moments that may not even seem at first glance related. Less, in that it is stripped of any accompanying emotion and conjecture—all elements that are deemed extraneous to clarity of thought—and made as objective as a nonscientific reality could ever be. The result: the crime as an object of strict scientific inquiry, to be approached by the principles of the scientific method. Its servant: the human mind.

What Is the Scientific Method of Thought?

When we think of the scientific method, we tend to think of an experimenter in his laboratory, probably holding a test tube and wearing a white coat, who follows a series of steps that runs something like this: make some observations about a phenomenon; create a hypothesis to explain those observations; design an experiment to test the hypothesis; run the experiment; see if the results match your expectations; rework your hypothesis if you must; lather, rinse, and repeat. Simple seeming enough. But how to go beyond that? Can we train our minds to work like that automatically, all the time?

Holmes recommends we start with the basics. As he says in our first meeting with him, "Before turning to those moral and mental aspects of the matter which present the greatest difficulties, let the enquirer begin by mastering more elementary problems." The scientific method begins with the most mundane seeming of things: observation. Before you even begin to ask the questions that will define the investigation of a crime, a scientific experiment, or a decision as apparently simple as whether or not to invite a certain friend to dinner, you must first explore the essential groundwork. It's not for nothing that Holmes calls the foundations of his inquiry "elementary." For, that is precisely what they are, the very basis of how something works and what makes it what it is.

And that is something that not even every scientist acknowledges outright, so ingrained is it in his way of thinking. When a physicist dreams up a new experiment or a biologist decides to test the properties of a newly isolated compound, he doesn't always realize that his specific question, his approach, his hypothesis, his very view of what he is doing would be impossible without the elemental knowledge at his disposal, that he has built up over the years. Indeed, he may have a hard time telling you from where exactly he got the idea for a study—and why he first thought it would make sense.

After World War II, physicist Richard Feynman was asked to serve on the State Curriculum Commission, to choose high school science textbooks for California. To his consternation, the texts appeared to leave students more confused than enlightened. Each book he examined was worse than the one prior. Finally, he came upon a promising beginning: a series of pictures, of a windup toy, an automobile, and a boy on a bicycle. Under each was a question: "What makes it go?" At last, he thought, something that was going to explain the basic science, starting with the fundamentals of mechanics (the toy), chemistry (the car), and biology (the boy). Alas, his elation was short lived. Where he thought to finally see explanation, real understanding, he found instead four words: "Energy makes it go." But what was *that*? Why did it make it go? How did it make it go? These questions weren't ever acknowledged, never mind answered. As Feynman put it, "That doesn't *mean* anything. . . . It's just a *word*!" Instead, he argued, "What they should have done is to look at the windup toy, see that there are springs inside, learn about springs, learn about wheels, and never mind 'energy.' Later on, when the children know something about how the toy actually works, they can discuss the more general principles of energy."

Feynman is one of the few who rarely took his knowledge base for granted, who always remembered the building blocks, the elements that lay underneath each question and each principle. And that is precisely what Holmes means when he tells us that we must begin with the basics, with such mundane problems that they might seem beneath our notice. How can you hypothesize, how can you make testable theories if you don't first know what and how to observe, if you don't first understand the fundamental nature of the problem at hand, down to its most basic elements? (The simplicity is deceptive, as you will learn in the next two chapters.)

The scientific method begins with a broad base of knowledge, an understanding of the facts and contours of the problem you are trying to tackle. In the case of Holmes in *A Study in Scarlet*, it's the mystery behind a murder in an abandoned house on Lauriston Gardens. In your case, it may be a decision whether or not to change careers. Whatever the specific issue, you must define and formulate it in your mind as specifically as possible—and then you must fill it in with past experience and present observation. (As Holmes admonishes Lestrade and Gregson when the two detectives fail to note a similarity between the murder being investigated and an earlier case, "There is nothing new under the sun. It has all been done before.")

Only then can you move to the hypothesis-generation point. This is the moment where the detective engages his imagination, generating possible lines of inquiry into the course of events, and not just sticking to the most obvious possibility—in *A Study in Scarlet*, for instance, *rache* need not be *Rachel* cut short, but could also signify the German for *revenge*—or where you might brainstorm possible scenarios that may arise from pursuing a new job direction. But you don't just start hypothesizing at random: all the potential scenarios and explanations come from that initial base of knowledge and observation.

Only then do you test. What does your hypothesis imply? At this point, Holmes will investigate all lines of inquiry, eliminating them one by one until the one that remains, however improbable, must be the truth. And you will run through career change scenarios and try to play out the implications to their logical, full conclusion. That, too, is manageable, as you will later learn.

But even then, you're not done. Times change. Circumstances change. That original knowledge base must always be updated. As our environment changes, we must never forget to revise and retest out hypotheses. The revolutionary can, if we're not careful, become the irrelevant. The thoughtful can become unthinking through our failure to keep engaging, challenging, pushing.

That, in a nutshell, is the scientific method: understand and frame the problem; observe; hypothesize (or imagine); test and deduce; and repeat. To follow Sherlock Holmes is to learn to apply that same approach not just to external clues, but to your every thought—and then turn it around and apply it to the every thought of every other person who may be involved, step by painstaking step.

When Holmes first lays out the theoretical principles behind his approach, he boils it down to one main idea: "How much an observant man might learn by an accurate and systematic examination of all that came his way." And that "all" includes each and every thought; in Holmes's world, there is no such thing as a thought that is taken at face value. As he notes, "From a drop of water, a logician could infer the possibility of an Atlantic or a Niagara without having seen or heard of one or the other." In other words, given our existing knowledge base, we can use observation to deduce meaning from an otherwise meaningless fact. For what kind of scientist is that who lacks the ability to imagine and hypothesize the new, the unknown, the as-of-yet untestable?

This is the scientific method at its most basic. Holmes goes a step further. He applies the same principle to human beings: a Holmesian disciple will, "on meeting a fellow-mortal, learn at a glance to distinguish the history of the man and the trade or profession to which he belongs. Puerile as such an exercise may seem, it sharpens the faculties of observation, and teaches one where to look and what to look for." Each observation, each exercise, each simple inference drawn from a simple fact will strengthen your ability to engage in ever-more-complex machinations. It will lay the groundwork for new habits of thinking that will make such observation second nature.

That is precisely what Holmes has taught himself—and can now teach us—to do. For, at its most basic, isn't that the detective's appeal? Not only can he solve the hardest of crimes, but he does so with an approach that seems, well, elementary when you get right down to it. This approach is based in science, in specific steps, in habits of thought that can be learned, cultivated, and applied.

That all sounds good in theory. But how do you even begin? It does seem like an awfully big hassle to always think scientifically, to always have to pay attention and break things down and observe and hypothesize and deduce and everything in between. Well, it both is and isn't. On the one hand, most of us have a long way to go. As we'll see, our minds aren't meant to think like Holmes by default. But on the other hand, new thought habits can be learned and applied. Our brains are remarkably adept at learning new ways of thinking—and our neural connections are remarkably flexible, even into old age. By following Holmes's thinking in the following pages, we will learn how to apply his methodology to our everyday lives, to be present and mindful and to treat each choice, each problem, each situation with the care it deserves. At first it will seem unnatural. But with time and practice it will come to be as second nature for us as it is for him.

Pitfalls of the Untrained Brain

One of the things that characterizes Holmes's thinking—and the scientific ideal—is a natural skepticism and inquisitiveness toward the world. Nothing is taken at face value. Everything is scrutinized and considered, and only then accepted (or not, as the case may be). Unfortunately, our minds are, in their default state, averse to such an approach. In order to think like Sherlock Holmes, we first need to overcome a sort of natural resistance that pervades the way we see the world.

Most psychologists now agree that our minds operate on a so-called two-system basis. One system is fast, intuitive, reactionary—a kind of constant fight-or-flight vigilance of the mind. It doesn't require much conscious thought or effort and functions as a sort of status quo autopilot. The other is slower, more deliberative, more thorough, more logical—but also much more cognitively costly. It likes to sit things out as long as it can and doesn't step in unless it thinks it absolutely necessary.

Because of the mental cost of that cool, reflective system, we spend most of our thinking time in the hot, reflexive system, basically ensuring that our natural observer state takes on the color of that system: automatic, intuitive (and not always rightly so), reactionary, quick to judge. As a matter of course, we go. Only when something really catches our attention or forces us to stop or otherwise jolts us do we begin to know, turning on the more thoughtful, reflective, cool sibling.

I'm going to give the systems monikers of my own: the Watson system and the Holmes system. You can guess which is which. Think of the Watson system as our naive selves, operating by the lazy thought habits—the ones that come most naturally, the so-called path of least resistance—that we've spent our whole lives acquiring. And think of the Holmes system as our aspirational selves, the selves that we'll be once we're done learning how to apply his method of thinking to our everyday lives—and in so doing break the habits of our Watson system once and for all.

When we think as a matter of course, our minds are preset to accept whatever it is that comes to them. First we believe, and only then do we question. Put differently, it's like our brains initially see the world as a true/false exam where the default answer is always *true*. And while it takes no effort whatsoever to remain in *true* mode, a switch of answer to *false* requires vigilance, time, and energy.

Psychologist Daniel Gilbert describes it this way: our brains *must* believe something in order to process it, if only for a split second. Imagine I tell you to think of pink elephants. You obviously know that pink elephants don't actually exist. But when you read the phrase, you just for a moment had to picture a pink elephant in your head. In order to realize that it couldn't exist, you had to believe for a second that it *did* exist. We understand and believe in the same instant. Benedict de Spinoza was the first to conceive of this necessity of acceptance for comprehension, and, writing a hundred years before Gilbert, William James explained the principle as "All propositions, whether attributive or existential, are believed through the very fact of being conceived." Only after the conception do we effortfully engage in disbelieving something—and, as Gilbert points out, that part of the process can be far from automatic.

In the case of the pink elephants the disconfirming process is simple. It takes next to no effort or time—although it still does take your brain more effort to process than it would if I said gray elephant, since counterfactual information requires that additional step of verification and disconfirmation that true information does not. But that's not always true: not everything is as glaring as a pink elephant. The more complicated a concept or idea, or the less obviously true or false (*There are no poisonous snakes in Maine.* True or false? Go! But even that can be factually verified. How about: *The death penalty is not as harsh a punishment as life imprisonment.* What now?), the more effort is required. And it doesn't take much for the process to be disrupted or to not occur altogether. If we decide that the statement sounds plausible enough as is (*sure; no poisonous snakes in Maine; why not?*), we are more likely than not to just let it go. Likewise, if we are busy, stressed, distracted, or otherwise depleted mentally, we may keep something marked as true without ever having taken the time to verify it—when faced with multiple demands, our mental capacity is simply too limited to be able to handle everything at once, and the verification process is one of the first things to go. When that happens, we are left with uncorrected beliefs, things that we will later recall as true when they are, in fact, false. (Are there poisonous snakes in Maine? Yes, as a matter of fact there are. But get asked in a year, and who knows if you will remember that or the opposite—especially if you were tired or distracted when reading this paragraph.)

What's more, not everything is as black and white—or as pink and white, as the case may be—as the elephant. And not everything that our intuition *says* is black and white is so in reality. It's awfully easy to get tripped up. In fact, not only do we believe everything we hear, at least initially, but even when we have been told explicitly that a statement is false *before we hear it*, we are likely to treat it as true. For instance, in something known as the correspondence bias (a concept we'll revisit in greater detail), we assume that what a person says is what that person actually believes—and we hold on to that assumption even if we've been told explicitly that it isn't so; we're even likely to judge the speaker in its light. Think back to the previous paragraph; do you think that what I wrote about the death penalty is my actual belief? You have no basis on which to answer that question—I haven't given you my opinion—and yet, chances are you've already answered it by taking my statement *as* my opinion. More disturbing still, even if we hear something denied—for example, *Joe has no links to the Mafia*—we may end up misremembering the statement as lacking the negator and end up believing that Joe *does* have Mafia links—and even if we don't, we are much more likely to form a negative opinion of Joe. We're even apt to recommend a longer prison sentence for him if we play the role of jury. Our tendency to confirm and to believe just a little too easily and often has very real consequences both for ourselves and for others.

Holmes's trick is to treat every thought, every experience, and every perception the way he would a pink elephant. In other words, begin with a healthy dose of skepticism instead of the credulity that is your mind's natural state of being. Don't just assume anything is the way it is. Think of everything as being as absurd as an animal that can't possibly exist in nature. It's a difficult proposition, especially to take on all at once—after all, it's the same thing as asking your brain to go from its natural resting state to a mode of constant physical activity, expending important energy even where it would normally yawn, say okay, and move on to the next thing—but not an impossible one, especially if you've got Sherlock Holmes on your side. For he, perhaps better than anyone else, can serve as a trusty companion, an ever-present model for how to accomplish what may look at first glance like a herculean task.

By observing Holmes in action, we will become better at observing our own minds. "How the deuce did he know that I had come from Afghanistan?" Watson asks Stamford, the man who has introduced him to Holmes for the first time.

Stamford smiles enigmatically in response. "That's just his little peculiarity," he tells Watson. "A good many people have wanted to know how he finds things out."

That answer only piques Watson's curiosity further. It's a curiosity that can only be satisfied over the course of long and detailed observation—which he promptly undertakes.

To Sherlock Holmes, the world has become by default a pink elephant world. It's a world where every single input is examined with the same care and healthy skepticism as the most absurd of animals. And by the end of this book, if you ask yourself the simple question, *What would Sherlock Holmes do and think in this situation?* you will find that your own world is on its way to being one, too. That thoughts that you never before realized existed are being stopped and questioned before being allowed to infiltrate your mind. That those same thoughts, properly filtered, can no longer slyly influence your behavior without your knowledge.

And just like a muscle that you never knew you had—one that suddenly begins to ache, then develop and bulk up as you begin to use it more and more in a new series of exercises—with practice your mind will see that the constant observation and never-ending scrutiny will become easier. (In fact, as you'll learn later in the book, it really is like a muscle.) It will become, as it is to Sherlock Holmes, second nature. You will begin to intuit, to deduce, to *think* as a matter of course, and you will find that you no longer have to give it much conscious effort.

Don't for a second think it's not doable. Holmes may be fictional, but Joseph Bell was very real. So, too, was Conan Doyle (and George Edalji wasn't the only beneficiary of his approach; Sir Arthur also worked to overturn the convictions of the falsely imprisoned Oscar Slater).

And maybe Sherlock Holmes so captures our minds for the very reason that he makes it seem possible, effortless even, to think in a way that would bring the average person to exhaustion. He makes the most rigorous scientific approach to thinking seem attainable. Not for nothing does Watson always exclaim, after Holmes gives him an explanation of his methods, that the thing couldn't have been any clearer. Unlike Watson, though, we can learn to see the clarity before the fact.

The Two Ms: Mindfulness and Motivation

It won't be easy. As Holmes reminds us, "Like all other arts, the Science of Deduction and Analysis is one which can only be acquired by long and patient study nor is life long enough to allow any mortal to attain the highest possible perfection in it." But it's also more than mere fancy. In essence, it comes down to one simple formula: to move from a System Watson– to a System Holmes–governed thinking takes mindfulness plus motivation. (That, and a lot of practice.) Mindfulness, in the sense of constant presence of mind, the attentiveness and hereness that is so essential for real, active observation of the world. Motivation, in the sense of active engagement and desire.

When we do such decidedly unremarkable things as misplacing our keys or losing our glasses only to find them on our head, System Watson is to blame: we go on a sort of autopilot and don't note our actions as we make them. It's why we often forget what we were doing if we're interrupted, why we stand in the middle of the kitchen wondering why we've entered it. System Holmes offers the type of retracing of steps that requires attentive recall, so that we break the autopilot and instead remember just where and why we did what we did. We aren't motivated or mindful all the time, and mostly it doesn't matter. We do things mindlessly to conserve our resources for something more important than the location of our keys.

But in order to break from that autopiloted mode, we have to be motivated to think in a mindful, present fashion, to exert effort on what goes through our heads instead of going with the flow. To think like Sherlock Holmes, we must *want*, actively, to think like him. In fact, motivation is so essential that researchers have often lamented the difficulty of getting accurate performance comparisons on cognitive tasks for older and younger participants. Why? The older adults are often far more motivated to perform well. They try harder. They engage more. They are more serious, more present, more involved. To them, the performance matters a great deal. It says something about their mental capabilities—and they are out to prove that they haven't lost the touch as they've aged. Not so younger adults. There is no comparable imperative. How, then, can you accurately compare the two groups? It's a question that continues to plague research into aging and cognitive function.

But that's not the only domain where it matters. Motivated subjects *always* outperform. Students who are motivated perform better on something as seemingly immutable as the IQ test—on average, as much as .064 standard deviations better, in fact. Not only that, but motivation predicts higher academic performance, fewer criminal convictions, and better employment outcomes. Children who have a so-called "rage to master"—a term coined by Ellen Winner to describe the intrinsic motivation to master a specific domain—are more likely to be successful in any number of endeavors, from art to science. If we are motivated to learn a language, we are more likely to succeed in our quest. Indeed, when we learn anything new, we learn better if we are motivated learners. Even our memory knows if we're motivated or not: we remember better if we were motivated at the time the memory was formed. It's called motivated encoding.

And then, of course, there is that final piece of the puzzle: practice, practice, practice. You have to supplement your mindful motivation with brutal training, thousands of hours of it. There is no way around it. Think of the phenomenon of expert knowledge: experts in all fields, from master chess players to master detectives, have superior memory in their field of choice. Holmes's knowledge of crime is ever at his fingertips. A chess player often holds hundreds of games, with all of their moves, in his head, ready for swift access. Psychologist K. Anders Ericsson argues that experts even see the world differently within their area of expertise: they see things that are invisible to a novice; they are able to discern patterns at a glance that are anything but obvious to an untrained eye; they see details as part of a whole and know at once what is crucial and what is incidental.

Even Holmes could not have begun life with System Holmes at the wheel. You can be sure that in his fictional world he was born, just as we are, with Watson at the controls. He just hasn't let himself stay that way. He took System Watson and taught it to operate by the rules of System Holmes, imposing reflective thought where there should rightly be reflexive reaction.

For the most part, System Watson is the habitual one. But if we are conscious of its power, we can ensure that it is not in control nearly as often as it otherwise would be. As Holmes often notes, he has made it a habit to engage his Holmes system, every moment of every day. In so doing, he has slowly trained his quick-to-judge inner Watson to perform as his public outer Holmes. Through sheer force of habit and will, he has taught his instant judgments to follow the train of thought of a far more reflective approach. And because this foundation is in place, it takes a matter of seconds for him to make his initial observations of Watson's character. That's why Holmes calls it intuition. Accurate intuition, the intuition that Holmes possesses, is of necessity based on training, hours and hours of it. An expert may not always realize consciously where it's coming from, but it comes from some habit, visible or not. What Holmes has done is to clarify the process, break down how hot can become cool, reflexive become reflective. It's what Anders Ericsson calls expert knowledge: an ability born from extended and intense practice and not some innate genius. It's not that Holmes was born to be the consulting detective to end all consulting detectives. It's that he has practiced his mindful approach to the world and has, over time, perfected his art to the level at which we find it.

As their first case together draws to a close, Dr. Watson compliments his new companion on his masterful accomplishment: "You have brought detection as near an exact science as it ever will be brought in this world." A high compliment indeed. But in the following pages, you will learn to do the exact same thing for your every thought, from its very inception— just as Arthur Conan Doyle did in his defense of George Edalji, and Joseph Bell in his patient diagnoses.

Sherlock Holmes came of age at a time when psychology was still in its infancy. We are far better equipped than he could have ever been. Let's learn to put that knowledge to good use.

SHERLOCK HOLMES FURTHER READING

"How the deuce did he know . . ." from *A Study in Scarlet*, chapter 1: Mr. Sherlock Holmes, p. 7.[1]
"Before turning to those moral and mental aspects . . ." *"How much an observant man might learn . . ."* *"Like all other arts, the Science of Deduction and Analysis . . ."* from *A Study in Scarlet*, chapter 2: The Science of Deduction, p. 15.

---

[1] All page numbers for this and subsequent "Further Reading" sections taken from editions specified at the end of the book.

CHAPTER TWO

The Brain Attic: What Is It and What's in There?

One of the most widely held notions about Sherlock Holmes has to do with his supposed ignorance of Copernican theory. "What the deuce is [the solar system] to me?" he exclaims to Watson in *A Study in Scarlet*. "You say that we go round the sun. If we went round the moon it would not make a pennyworth of difference to me or to my work." And now that he knows that fact? "I shall do my best to forget it," he promises.

It's fun to home in on that incongruity between the superhuman-seeming detective and a failure to grasp a fact so rudimentary that even a child would know it. And ignorance of the solar system is quite an omission for someone who we might hold up as the model of the scientific method, is it not? Even the BBC series *Sherlock* can't help but use it as a focal point of one of its episodes.

But two things about that perception bear further mention. First, it isn't, strictly speaking, true. Witness Holmes's repeated references to astronomy in future stories—in "The Musgrave Ritual," he talks about "allowances for personal equation, as the astronomers would have it"; in "The Greek Interpreter," about the "obliquity of the ecliptic"; in "The Adventure of the Bruce-Partington Plans," about "a planet leaving its orbit." Indeed, eventually Holmes does use almost all of the knowledge that he denies having at the earliest stages of his friendship with Dr. Watson. (And in true-to-canon form, *Sherlock* the BBC series does end on a note of scientific triumph: Holmes does know astronomy after all, and that knowledge saves the day—and the life of a little boy.)

In fact, I would argue that he exaggerates his ignorance precisely to draw our attention to a second—and, I think, much more important— point. His supposed refusal to commit the solar system to memory serves to illustrate an analogy for the human mind that will prove to be central to Holmes's thinking and to our ability to emulate his methodology. As Holmes tells Watson, moments after the Copernican incident, "I consider that a man's brain originally is like a little empty attic, and you have to stock it with such furniture as you choose."

When I first heard the term *brain attic*—back in the days of firelight and the old crimson hardcover—all I could picture in my seven-year-old head was the cover of the black-and-white Shel Silverstein book that sat prominently on my bookshelf, with its half-smiling, lopsided face whose forehead was distended to a wrinkled triangle, complete with roof, chimney, and window with open shutters. Behind the shutters, a tiny face peeking out at the world. Is this what Holmes meant? A small room with sloped sides and a foreign creature with a funny face waiting to pull the cord and turn the light off or on?

As it turns out, I wasn't far from wrong. For Sherlock Holmes, a person's brain attic really is an incredibly concrete, physical space. Maybe it has a chimney. Maybe it doesn't. But whatever it looks like, it is a space in your head, specially fashioned for storing the most disparate of objects. And yes, there is certainly a cord that you can pull to turn the light on or off at will. As Holmes explains to Watson, "A fool takes in all the lumber of every sort that he comes across, so that the knowledge which might be useful to him gets crowded out, or at best is jumbled up with a lot of other things, so that he has a difficulty in laying his hands upon it. Now the skillful workman is very careful indeed as to what he takes into his brain-attic."

That comparison, as it turns out, is remarkably accurate. Subsequent research on memory formation, retention, and retrieval has—as you'll soon see—proven itself to be highly amenable to the attic analogy. In the chapters that follow, we will trace the role of the brain attic from the inception to the culmination of the thought process, exploring how its structure and content work at every point—and what we can do to improve that working on a regular basis.

The attic can be broken down, roughly speaking, into two components: structure and contents. The attic's structure is how our mind works: how it takes in information. How it processes that information. How it sorts it and stores it for the future. How it may choose to integrate it or not with contents that are already in the attic space. Unlike a physical attic, the structure of the brain attic isn't altogether fixed. It can expand, albeit not indefinitely, or it can contract, depending on how we use it (in other words, our memory and processing can become more or less effective). It can change its mode of retrieval (*How do I recover information I've stored?*). It can change its storage system (*How do I deposit information I've taken in: where will it go? how will it be marked? how will it be integrated?*). At the end, it will have to remain within certain confines—each attic, once again, is different and subject to its unique constraints—but within those confines, it can take on any number of configurations, depending on how we learn to approach it.

The attic's contents, on the other hand, are those things that we've taken in from the world and that we've experienced in our lives. Our memories. Our past. The base of our knowledge, the information we start with every time we face a challenge. And just like a physical attic's contents can change over time, so too does our mind attic continue to take in and discard items until the very end. As our thought process begins, the furniture of memory combines with the structure of internal habits and external circumstances to determine which item will be retrieved from storage at any given point. Guessing at the contents of a person's attic from his outward appearance becomes one of Sherlock's surest ways of determining who that person is and what he is capable of.

As we've already seen, much of the original intake is outside of our control: just like we must picture a pink elephant to realize one doesn't exist, we can't help but become acquainted—if only for the briefest of moments—with the workings of the solar system or the writings of Thomas Carlyle should Watson choose to mention them to us. We can, however, learn to master many aspects of our attic's structure, throwing out junk that got in by mistake (as Holmes promises to forget Copernicus at the earliest opportunity), prioritizing those things we want to and pushing back those that we don't, learning how to take the contours of our unique attic into account so that they don't unduly influence us as they otherwise might.

While we may never become quite as adept as the master at divining a man's innermost thoughts from his exterior, in learning to understand the layout and functionality of our own brain attic we take the first step to becoming better at exploiting its features to their maximum potential— in other words, to learning how to optimize our own thought process, so that we start any given decision or action as our best, most aware selves. Our attic's structure and contents aren't there because we *have* to think that way, but because we've learned over time and with repeat practice (often unknown, but practice nevertheless) to think that way. We've decided, on a certain level, that mindful attention is just not worth the effort. We've chosen efficiency over depth. It may take just as long, but we can learn to think differently.

The basic structure may be there for good, but we can learn to alter its exact linkages and building blocks—and that alteration will actually rebuild the attic, so to speak, rewiring our neural connections as we change our habits of thought. Just as with any renovation, some of the major overhauls may take some time. You can't just rebuild an attic in a day. But some minor changes will likely begin to appear within days—and even hours. And they will do so no matter how old your attic is and how long it has been since it's gotten a proper cleaning. In other words, our brains can learn new skills quickly—and they can continue to do so throughout our lives, not just when we are younger. As for the contents: while some of those, too, are there to stay, we can be selective about what we keep in the future—and can learn to organize the attic so that those contents we do want are easiest to access, and those we either value less or want to avoid altogether move further into the corners. We may not come out with an altogether different attic, but we can certainly come out with one that more resembles Holmes's.

Memory's Furniture

The same day that Watson first learns of his new friend's theories on deduction—all of that Niagara-from-a-drop-of-water and whatnot—he is presented with a most convincing demonstration of their power: their application to a puzzling murder. As the two men sit discussing Holmes's article, they are interrupted by a message from Scotland Yard. Inspector

Tobias Gregson requests Holmes's opinion on a puzzler of a case. A man has been found dead, and yet, "There had been no robbery, nor is there any evidence as to how the man met his death. There are marks of blood in the room, but there is no wound upon his person." Gregson continues his appeal: "We are at a loss as to how he came into the empty house; indeed, the whole affair is a puzzler." And without further ado, Holmes departs for Lauriston Gardens, Watson at his side.

Is the case as singular as all that? Gregson and his colleague, Inspector Lestrade, seem to think so. "It beats anything I have seen, and I am no chicken," offers Lestrade. Not a clue in sight. Holmes, however, has an idea. "Of course, this blood belongs to a second individual—presumably the murderer, if murder has been committed," he tells the two policemen. "It reminds me of the circumstances attendant on the death of Van Jansen, in Utrecht, in the year '34. Do you remember the case, Gregson?"

Gregson confesses that he does not.

"Read it up—you really should," offers Holmes. "There is nothing new under the sun. It has all been done before."

Why does Holmes remember Van Jansen while Gregson does not? Presumably, both men had at one point been acquainted with the circumstances—after all, Gregson has had to train extensively for his current position—and yet the one has retained them for his use, while for the other they have evaporated into nonexistence.

It all has to do with the nature of the brain attic. Our default System Watson attic is jumbled and largely mindless. Gregson may have once known about Van Jansen but has lacked the requisite motivation and presence to retain his knowledge. Why should he care about old cases? Holmes, however, makes a conscious, motivated choice to remember cases past; one never knows when they might come in handy. In his attic, knowledge does not get lost. He has made a deliberate decision that these details matter. And that decision has, in turn, affected how and what—and when—he remembers.

Our memory is in large part the starting point for how we think, how our preferences form, and how we make decisions. It is the attic's content that distinguishes even an otherwise identically structured mind from its neighbor's. What Holmes means when he talks about stocking your attic with the appropriate furniture is the need to carefully choose which experiences, which memories, which aspects of your life you want to hold on to beyond the moment when they occur. (He should know: he would not have even existed as we know him had Arthur Conan Doyle not retrieved his experiences with Dr. Joseph Bell from memory in creating his fictional detective.) He means that for a police inspector, it would be well to remember past cases, even seemingly obscure ones: aren't they, in a sense, the most basic knowledge of his profession?

In the earliest days of research, memory was thought to be populated with so-called engrams, memory traces that were localized in specific parts of the brain. To locate one such engram—for the memory of a maze—psychologist Karl Lashley taught rats to run through a labyrinth. He then cut out various parts of their brain tissue and put them right back into the maze. Though the rats' motor function declined and some had to hobble or crawl their way woozily through the twists and turns, the animals never altogether forgot their way, leading Lashley to conclude that there was no single location that stored a given memory. Rather, memory was widely distributed in a connected neural network—one that may look rather familiar to Holmes.

Today, it is commonly accepted that memory is divided into two systems, one short- and one long-term, and while the precise mechanisms of the systems remain theoretical, an atticlike view—albeit a very specific kind of attic—may not be far from the truth. When we see something, it is first encoded by the brain and then stored in the hippocampus—think of it as the attic's first entry point, where you place everything before you know whether or not you will need to retrieve it. From there, the stuff that you either actively consider important or that your mind somehow decides is worth storing, based on past experience and your past directives (i.e., what you normally consider important), will be moved to a specific box within the attic, into a specific folder, in a specific compartment in the cortex—the bulk of your attic's storage space, your long-term memory. This is called consolidation. When you need to recall a specific memory that has been stored, your mind goes to the proper file and pulls it out. Sometimes it pulls out the file next to it, too, activating the contents of the whole box or whatever happens to be nearby—associative activation. Sometimes the file slips and by the time you get it out into the light, its contents have changed from when you first placed them inside— only you may not be aware of the change. In any case, you take a look, and you add anything that may seem newly relevant. Then you replace it in its spot in its changed form. Those steps are called retrieval and recon- solidation, respectively.

The specifics aren't nearly as important as the broad idea. Some things get stored; some are thrown out and never reach the main attic. What's stored is organized according to some associative system—your brain de- cides where a given memory might fit—but if you think you'll be retriev- ing an exact replica of what you've stored, you're wrong. Contents shift, change, and re-form with every shake of the box where they are stored. Put in your favorite book from childhood, and if you're not careful, the next time you retrieve it there may be water damage to the picture you so wanted to see. Throw a few photo albums up there, and the pictures may get mixed together so that the images from one trip merge with those from another one altogether. Reach for an object more often, and it doesn't gather dust. It stays on top, fresh and ready for your next touch (though who knows what it may take with it on its next trip out). Leave it un- touched, and it retreats further and further into a heap—but it can be dislodged by a sudden movement in its vicinity. Forget about something for long enough, and by the time you go to look for it, it may be lost be- yond your reach—still there, to be sure, but at the bottom of a box in a dark corner where you aren't likely to ever again find it.

To cultivate our knowledge actively, we need to realize that items are being pushed into our attic space at every opportunity. In our default state, we don't often pay attention to them unless some aspect draws our attention—but that doesn't mean they haven't found their way into our attic all the same. They sneak in if we're not careful, if we just passively take in information and don't make a conscious effort to control our at- tention (something we'll learn about a bit further on)—especially if they are things that somehow pique our attention naturally: topics of general interest; things we can't help but notice; things that raise some emotion in us; or things that capture us by some aspect of novelty or note.

It is all too easy to let the world come unfiltered into your attic space, populating it with whatever inputs may come its way or whatever naturally captures your attention by virtue of its interest or immediate relevance to you. When we're in our default System Watson mode, we don't "choose" which memories to store. They just kind of store themselves—or they don't, as the case may be. Have you ever found yourself reliving a memory with a friend—that time you both ordered the ice cream sundae instead of lunch and then spent the afternoon walking around the town center and people-watching by the river—only to find that the friend has no idea what you're talking about? *It must have been someone else,* he says. *Not me. I'm not a sundae type of guy.* Only, you know it was him. Conversely, have you ever been on the receiving end of that story, having someone recount an experience or event or moment that you simply have no recollection of? And you can bet that that someone is just as certain as you were that it happened just the way he recalls.

But that, warns Holmes, is a dangerous policy. Before you know it, your mind will be filled with so much useless junk that even the information that happened to be useful is buried so deeply and is so inaccessible that it might as well not even be there. It's important to keep one thing in mind: we know only what we can remember at any given point. In other words, no amount of knowledge will save us if we can't recall it at the moment we need it. It doesn't matter if the modern Holmes knows anything about astronomy if he can't remember the timing of the asteroid that appears in a certain painting at the crucial moment. A boy will die and Benedict Cumberbatch will upset our expectations. It doesn't matter if Gregson once knew of Van Jansen and all his Utrecht adventures. If he can't remember them at Lauriston Gardens, they do him no good whatsoever.

When we try to recall something, we won't be able to do so if there is too much piled up in the way. Instead, competing memories will vie for our attention. I may try to remember that crucial asteroid and think instead of an evening where I saw a shooting star or what my astronomy professor was wearing when she first lectured to us about comets. It all depends on how well organized my attic is—how I encoded the memory to begin with, what cues are prompting its retrieval now, how methodical and organized my thought process is from start to finish. I may have stored something in my attic, but whether or not I have done so accurately and in a way that can be accessed in a timely fashion is another question altogether. It's not as simple as getting one discrete item out whenever I want it just because I once stuffed it up there.

But that need not be the case. Inevitably, junk will creep into the attic. It's impossible to be as perfectly vigilant as Holmes makes himself out to be. (You'll learn later that he isn't quite as strict, either. Useless junk may end up being flea market gold in the right set of circumstances.) But it *is* possible to assert more control over the memories that do get encoded.

If Watson—or Gregson, as the case may be—wanted to follow Holmes's method, he would do well to realize the motivated nature of encoding: we remember more when we are *interested* and *motivated*. Chances are, Watson was quite capable of retaining his medical training—and the minutiae of his romantic escapades. These were things that were relevant to him and captured his attention. In other words, he was motivated to remember.

Psychologist Karim Kassam calls it the Scooter Libby effect: during his 2007 trial, Lewis "Scooter" Libby claimed no memory of having mentioned the identity of a certain CIA employee to any reporters of government officials. The jurors didn't buy it. How could he not remember something so important? Simple. It wasn't nearly as important at the time as it was in retrospect—and where motivation matters most is at the moment we are storing memories in our attics to begin with, and not afterward. The so-called Motivation to Remember (MTR) is far more important at the point of encoding—and no amount of MTR at retrieval will be efficient if the information wasn't properly stored to begin with. As hard as it is to believe, Libby may well have been telling the truth.

We can take advantage of MTR by activating the same processes consciously when we need them. When we really want to remember something, we can make a point of paying attention to it, of saying to ourselves, *This, I want to remember*—and, if possible, solidifying it as soon as we can, whether it be by describing an experience to someone else or to ourselves, if no one else is available (in essence, rehearsing it to help consolidation). Manipulating information, playing around with it and talking it through, making it come alive through stories and gestures, may be much more effective in getting it to the attic when you want it to get there than just trying to think it over and over. In one study, for instance, students who explained mathematical material after reading it once did better on a later test than those who repeated that material several times. What's more, the more cues we have, the better the likelihood of successful retrieval. Had Gregson originally focused on all of the Utrecht details at the moment he first learned of the case—sights, smells, sounds, whatever else was in the paper that day—and had he puzzled over the case in various guises, he would be far more likely to recall it now. Likewise, had he linked it to his existing knowledge base—in other words, instead of moving a fresh box or folder into his attic, had he integrated it into an existing, related one, be it on the topic of bloody crime scenes with bloodless bodies, or cases from 1834, or whatever else—the association would later facilitate a prompt response to Holmes's question. Anything to distinguish it and make it somehow more personal, relatable, and—crucially—memorable. Holmes remembers the details that matter to him—and not those that don't. At any given moment, you only think you know what you know. But what you really know is what you can recall.

So what determines what we can and can't remember at a specific point in time? How is the content of our attic activated by its structure?

The Color of Bias: The Attic's Default Structure

It is autumn 1888, and Sherlock Holmes is bored. For months, no case of note has crossed his path. And so the detective takes solace, to Dr. Watson's great dismay, in the 7 percent solution: cocaine. According to Holmes, it stimulates and clarifies his mind—a necessity when no food for thought is otherwise available.

"Count the cost!" Watson tries to reason with his flatmate. "Your brain may, as you say, be roused and excited, but it is a pathological and morbid process which involves increased tissue-change and may at least leave a permanent weakness. You know, too, what a black reaction comes upon you. Surely the game is hardly worth the candle."

Holmes remains unconvinced. "Give me problems, give me work, give me the most abstruse cryptogram, or the most intricate analysis," he says, "and I am in my own proper atmosphere. I can dispense then with artificial stimulant. But I abhor the dull routine of existence." And none of Dr. Watson's best medical arguments will make a jot of difference (at least not for now).

Luckily, however, in this particular instance they don't need to. A crisp knock on the door, and the men's landlady, Mrs. Hudson, enters with an announcement: a young lady by the name of Miss Mary Morstan has arrived to see Sherlock Holmes. Watson describes Mary's entrance:

> Miss Morstan entered the room with a firm step and an outward composure of manner. She was a blonde young lady, small, dainty, well gloved, and dressed in the most perfect taste. There was, however, a plainness and simplicity about her costume which bore with it a suggestion of limited means. The dress was a sombre grayish beige, untrimmed and unbraided, and she wore a small turban of the same dull hue, relieved only by a suspicion of white feather in the side. Her face had neither regularity of feature nor beauty of complexion, but her expression was sweet and amiable, and her large blue eyes were singularly spiritual and sympathetic. In an experience of women which extends over many nations and three separate continents, I have never looked upon a face which gave a clearer promise of a refined and sensitive nature. I could not but observe that as she took the seat which Sherlock Holmes placed for her, her lip trembled, her hand quivered, and she showed every sign of intense inward agitation.

Who might this lady be? And what could she want with the detective? These questions form the starting point of *The Sign of Four*, an adventure that will take Holmes and Watson to India and the Andaman Islands, pygmies and men with wooden legs. But before any of that there is the lady herself: who she is, what she represents, where she will lead. In a few pages, we will examine the first encounter between Mary, Holmes, and Watson and contrast the two very different ways in which the men react to their visitor. But first, let's take a step back to consider what happens in our mind attic when we first enter a situation—or, as in the case of *The Sign of Four*, encounter a person. How do those contents that we've just examined actually become activated?

From the very first, our thinking is governed by our attic's so-called structure: its habitual modes of thought and operation, the way in which we've learned, over time, to look at and evaluate the world, the biases and heuristics that shape our intuitive, immediate perception of reality. Though, as we've just seen, the memories and experiences stored in an individual attic vary greatly from person to person, the general patterns of activation and retrieval remain remarkably similar, coloring our thought process in a predictable, characteristic fashion. And if these habitual patterns point to one thing, it's this: our minds love nothing more than jumping to conclusions.

Imagine for a moment that you're at a party. You're standing in a group of friends and acquaintances, chatting happily away, drink in hand, when you glimpse a stranger angling his way into the conversation. By the time he has opened his mouth—even before he has even quite made it to the group's periphery—you have doubtless already formed any number of preliminary impressions, creating a fairly complete, albeit potentially inaccurate, picture of who this stranger is as a person. How is Joe Stranger dressed? Is he wearing a baseball hat? You love (hate) baseball. This must be a great (boring) guy. How does he walk and hold himself? What does he look like? Oh, is he starting to bald? What a downer. Does he actually think he can hang with someone as young and hip as you? What does he seem like? You've likely assessed how similar or different he is from you— same gender? race? social background? economic means?—and have even assigned him a preliminary personality—shy? outgoing? nervous? self-confident?—based on his appearance and demeanor alone. Or, maybe Joe Stranger is actually Jane Stranger and her hair is dyed the same shade of blue as your childhood best friend dyed *her* hair right before you stopped talking to each other, and you always thought the hair was the first sign of your impending break, and now all of a sudden, all of these memories are clogging your brain and coloring the way you see this new person, innocent Jane. You don't even notice anything else.

As Joe or Jane start talking, you'll fill in the details, perhaps rearranging some, amplifying others, even deleting a few entirely. But you'll hardly ever alter your initial impression, the one that started to form the second Joe or Jane walked your way. And yet what is that impression based on? Is it really anything of substance? You only happened to remember your ex–best friend, for instance, because of an errant streak of hair.

When we see Joe or Jane, each question we ask ourselves and each detail that filters into our minds, floating, so to speak, through the little attic window, primes our minds by activating specific associations. And those associations cause us to form a judgment about someone we have never even met, let alone spoken to.

You may want to hold yourself above such prejudices, but consider this. The Implicit Association Test (IAT) measures the distance between your conscious attitudes—those you are aware of holding—and your unconscious ones—those that form the invisible framework of your attic, beyond your immediate awareness. The measure can test for implicit bias toward any number of groups (though the most common one tests racial biases) by looking at reaction times for associations between positive and negative attributes and pictures of group representatives. Sometimes the stereotypical positives are represented by the same key: "European American" and "good," for instance, are both associated with, say, the "I" key, and "African American" and "bad" with the "E" key. Sometimes they are represented by different ones: now, the "I" is for "African American" and "good," while "European American" has moved to the "bad," "E" key. Your speed of categorization in each of these circumstances determines your implicit bias. To take the racial example, if you are faster to categorize when "European American" and "good" share a key and "African American" and "bad" share a key, it is taken as evidence of an implicit race bias.[2]

The findings are robust and replicated extensively: even those individuals who score the absolute lowest on self-reported measures of stereotype attitudes (for example, on a four-point scale ranging from Strongly

[2] You can take the IAT yourself online, at Harvard University's "Project Implicit" website, implicit.harvard.edu.

Female to Strongly Male, do you most strongly associate career with male or female?) often show a difference in reaction time on the IAT that tells a different story. On the race-related attitudes IAT, about 68 percent of over 2.5 million participants show a biased pattern. On age (i.e., those who prefer young people over old): 80 percent. On disability (i.e., those who favor people *without* any disabilities): 76 percent. On sexual orientation (i.e., those who favor straight people over gay): 68 percent. On weight (i.e., those who favor thin people over fat): 69 percent. The list goes on and on. And those biases, in turn, affect our decision making. How we see the world to begin with will impact what conclusions we reach, what evaluations we form, and what choices we make at any given point.

This is not to say that we will necessarily act in a biased fashion; we are perfectly capable of resisting our brains' basic impulses. But it does mean that the biases are there at a very fundamental level. Protest as you may that it's just not you, but more likely than not, it is. Hardly anyone is immune altogether.

Our brains are wired for quick judgments, equipped with back roads and shortcuts that simplify the task of taking in and evaluating the countless inputs that our environment throws at us every second. It's only natural. If we truly contemplated every element, we'd be lost. We'd be stuck. We'd never be able to move beyond that first evaluative judgment. In fact, we may not be able to make any judgment at all. Our world would become far too complex far too quickly. As William James put it, "If we remembered everything, we should on most occasions be as ill off as if we remembered nothing."

Our way of looking at and thinking about the world is tough to change and our biases are remarkably sticky. But tough and sticky doesn't mean unchangeable and immutable. Even the IAT, as it turns out, can be bested—after interventions and mental exercises that target the very biases it tests, that is. For instance, if you show individuals pictures of blacks enjoying a picnic before you have them take the racial IAT, the bias score decreases significantly.

A Holmes and a Watson may both make instantaneous judgments—but the shortcuts their brains are using could not be more different. Whereas Watson epitomizes the default brain, the structure of our mind's connections in their usual, largely passive state, Holmes shows what is possible: how we can rewire that structure to circumvent those instantaneous reactions that prevent a more objective and thorough judgment of our surroundings.

For instance, consider the use of the IAT in a study of medical bias. First, each doctor was shown a picture of a fifty-year-old man. In some pictures, the man was white. In some, he was black. The physicians were then asked to imagine the man in the picture as a patient who presented with symptoms that resembled a heart attack. How would they treat him? Once they gave an answer, they took the racial IAT.

In one regard, the results were typical. Most doctors showed some degree of bias on the IAT. But then, an interesting thing happened: bias on the test did not necessarily translate into bias in treating the hypothetical patient. On average, doctors were just as likely to say they would prescribe the necessary drugs to blacks as to whites—and oddly enough, the more seemingly biased physicians actually treated the two groups more equally than the less biased ones.

What our brains do on the level of instinct and how we act are not one and the same. Does this mean that biases disappeared, that their brains didn't leap to conclusions from implicit associations that occurred at the most basic level of cognition? Hardly. But it does mean that the right motivation can counteract such bias and render it beside the point in terms of actual behavior. How our brains jump to conclusions is not how we are destined to act. Ultimately, our behavior is ours to control—if only we want to do so.

What happened when you saw Joe Stranger at the cocktail party is the exact same thing that happens even to someone as adept at observation as Mr. Sherlock Holmes. But just like the doctors who have learned over time to judge based on certain symptoms and disregard others as irrelevant, Holmes has learned to filter his brain's instincts into those that should and those that should not play into his assessment of an unknown individual.

What enables Holmes to do this? To observe the process in action, let's revisit that initial encounter in *The Sign of Four*, when Mary Morstan, the mysterious lady caller, first makes her appearance. Do the two men see Mary in the same light? Not at all. The first thing Watson notices is the lady's appearance. She is, he remarks, a rather attractive woman. Irrelevant, counters Holmes. "It is of the first importance not to allow your judgment to be biased by personal qualities," he explains. "A client is to me a mere unit, a factor in a problem. The emotional qualities are antagonistic to clear reasoning. I assure you that the most winning woman I ever knew was hanged for poisoning three children for their insurance-money, and the most repellent man of my acquaintance is a philanthropist who has spent nearly a quarter of a million upon the London poor."

But Watson won't have it. "In this case, however—" he interrupts.

Holmes shakes his head. "I never make exceptions. An exception disproves the rule."

Holmes's point is clear enough. It's not that you won't experience emotion. Nor are you likely to be able to suspend the impressions that form almost automatically in your mind. (Of Miss Morstan, he remarks, "I think she is one of the most charming young ladies I ever met"—as high a compliment from Holmes as they come.) But you don't have to let those impressions get in the way of objective reasoning. ("But love is an emotional thing, and whatever is emotional is opposed to that true cold reason which I place above all things," Holmes immediately adds to his acknowledgment of Mary's charm.) You can recognize their presence, and then consciously cast them aside. You can acknowledge that Jane reminds you of your high school frenemy, and then move past it. That emotional luggage doesn't matter nearly as much as you may think it does. And never think that something is an exception. It's not.

But oh how difficult it can be to apply either of these principles—the discounting of emotion or the need to never make exceptions, no matter how much you may want to—in reality. Watson desperately wants to believe the best about the woman who so captivates him, and to attribute anything unfavorable about her to less-than-favorable circumstances. His undisciplined mind proceeds to violate each of Holmes's rules for proper reasoning and perception: from making an exception, to allowing in emotion, to failing altogether to attain that cold impartiality that Holmes makes his mantra.

From the very start, Watson is predisposed to think well of their guest. After all, he is already in a relaxed, happy mood, bantering in typical fashion with his detective flatmate. And rightly or wrongly, that mood will spill over into his judgment. It's called the affect heuristic: how we feel is how we think. A happy and relaxed state makes for a more accepting and less guarded worldview. Before Watson even knows that someone is soon to arrive, he is already set to like the visitor.

And once the visitor enters? It's just like that party. When we see a stranger, our mind experiences a predictable pattern of activation, which has been predetermined by our past experiences and our current goals—which includes our motivation—and state of being. When Miss Mary Morstan enters 221B Baker Street, Watson sees, "a blonde young lady, small, dainty, well gloved, and dressed in the most perfect taste. There was, however, a plainness and simplicity about her costume which bore with it a suggestion of limited means." Right away, the image stirs up memories in his head of other young, dainty blondes Watson has known—but not frivolous ones, mind you; ones who are plain and simple and undemanding, who do not throw their beauty in your face but smooth it over with a dress that is somber beige, "untrimmed and unbraided." And so, Mary's expression becomes "sweet and amiable, her large blue eyes were singularly spiritual and sympathetic." Watson concludes his opening paean with the words, "In an experience of women which extends over many nations and three separate continents, I have never looked upon a face which gave a clearer promise of a refined and sensitive nature."

Right away, the good doctor has jumped from a color of hair and complexion and a style of dress to a far more reaching character judgment. Mary's appearance suggests simplicity; perhaps so. But sweetness? Amiability? Spirituality? Sympathy? Refinement and sensitivity? Watson has no basis whatsoever for any of these judgments. Mary has yet to say a single word in his presence. All she has done is enter the room. But already a host of biases are at play, vying with one another to create a complete picture of this stranger.

In one moment, Watson has called on his reputedly vast experience, on the immense stores of his attic that are labeled WOMEN I'VE KNOWN, to flesh out his new acquaintance. While his knowledge of women may indeed span three separate continents, we have no reason to believe that his assessment here is accurate—unless, of course, we are told that in the past, Watson has always judged a woman's character successfully from first glance. And somehow I doubt that's the case. Watson is conveniently forgetting how long it took to get to know his past companions—assuming he ever got to know them at all. (Consider also that Watson is a bachelor, just returned from war, wounded, and largely friendless. What would his chronic motivational state likely be? Now, imagine he'd been instead married, successful, the toast of the town. Replay his evaluation of Mary accordingly.)

This tendency is a common and powerful one, known as the availability heuristic: we use what is available to the mind at any given point in time. And the easier it is to recall, the more confident we are in its applicability and truth. In one of the classic demonstrations of the effect, individuals who had read unfamiliar names in the context of a passage later judged those names as famous—based simply on the ease with which they could recall them—and were subsequently more confident in the accuracy of their judgments. To them, the ease of familiarity was proof enough. They didn't stop to think that availability based on earlier exposure could possibly be the culprit for their feelings of effortlessness.

Over and over, experimenters have demonstrated that when something in the environment, be it an image or a person or a word, serves as a prime, individuals are better able to access related concepts—in other words, those concepts have become more available—and they are more likely to use those concepts as confident answers, whether or not they are accurate. Mary's looks have triggered a memory cascade of associations in Watson's brain, which in turn creates a mental picture of Mary that is composed of whatever associations she happened to have activated but does not necessarily resemble the "real Mary." The closer Mary fits with the images that have been called up—the representativeness heuristic—the stronger the impression will be, and the more confident Watson will be in his objectivity.

Forget everything else that Watson may or may not know. Additional information is not welcome. Here's one question the gallant doctor isn't likely to ask himself: how many actual women does he meet who end up being refined, sensitive, spiritual, sympathetic, sweet, and amiable, all at once?

How typical is this type of person if you consider the population at large? Not very, I venture to guess—even if we factor in the blond hair and blue eyes, which are doubtless signs of saintliness and all. And how many women in total is he calling to mind when he sees Mary? One? Two? One hundred? What is the total sample size? Again, I'm willing to bet it is not very large— and the sample that has been selected is inherently a biased one.

While we don't know what precise associations are triggered in the doctor's head when he first sees Miss Morstan, my bet would be on the most recent ones (the recency effect), the most salient ones (the ones that are most colorful and memorable; all of those blue-eyed blondes who ended up being uninteresting, drab, and unimpressive? I doubt he is now remembering them; they may as well have never existed), and the most familiar ones (the ones that his mind has returned to most often—again, likely not the most representative of the lot). And those have biased his view of Mary from the onset. Chances are, from this point forward, it will take an earthquake, and perhaps even more than that, to shake Watson from his initial assessment.

His steadfastness will be all the stronger because of the physical nature of the initial trigger: faces are perhaps the most powerful cue we have—and the most likely to prompt associations and actions that just won't go away.

To see the power of the face in action, look at these pictures.

1. Which face is the more attractive? and 2. Which person is the more competent?

If I were to flash these pictures at you for as little as one-tenth of a second, your opinion would already most likely agree with the judgments of hundreds of others to whom I've shown pictures of these two individuals in the same way. But that's not all: those faces you just looked at aren't random. They are the faces of two rival political candidates, who ran in the 2004 U.S. senate election in Wisconsin. And the rating you gave for competence (an index of both strength and trustworthiness) will be highly predictive of the actual winner (it's the man on the left; did your competence evaluation match up?). In approximately 70 percent of cases, competence ratings given in under a second of exposure will predict the actual results of political races. And that predictability will hold in elections that range from the United States to England, from Finland to Mexico, and from Germany to Australia. From the strength of a chin and the trace of a smile, our brains decide who will serve us best. (And look at the result: Warren G. Harding, the most perfect square-jawed president that ever was.) We are wired to do just what we shouldn't: jump to conclusions based on some subtle, subconscious cue that we're not even aware of—and the repercussions extend to situations far more serious than Watson's trusting too much in a client's pretty face. Unprepared, he never stands a chance at that "true cold reason" that Holmes seems to hold in the tips of his fingers.

Just as a fleeting impression of competence can form the basis of a political vote, so Watson's initial overwhelmingly positive assessment of Mary lays the foundation for further action that reinforces that initial view. His judgments from here on out will be influenced strongly by the effects of primacy—the persistent strength of first impressions.

With his eyes shaded by a rosy glow, Watson is now much more likely to fall prey to the halo effect (if one element—here, physical appearance—strikes you as positive, you are likely to see the other elements as positive as well, and everything that doesn't fit will easily—and subconsciously—be reasoned away). He will also be susceptible to the classic correspondence bias: everything negative about Mary will be seen as a result of external circumstances—stress, strain, bad luck, whatever it may be— and everything positive of her character. She will get credit for all that's good, and the environment will shoulder blame for all that's bad. Chance and luck? Not important. The knowledge that we are, as a general rule, extremely bad at making *any* sort of prediction about the future, be it for an event or a behavior? Likewise irrelevant to his judgment. In fact, unlike Holmes, he likely hasn't even considered that possibility—or evaluated his own competence.

All the while, Watson will likely remain completely unaware of the hoops through which his mind is jumping to maintain a coherent impression of Mary, to form a narrative based on discrete inputs that makes sense and tells an intuitively appealing story. And in a self-fulfilling prophecy of sorts, which could potentially have rather perverse consequences, his own behavior could prompt Mary to act in a way that seems to confirm his initial impression of her. Act toward Mary as if she were a beautiful saint, and she will likely respond to him with a saintly smile. Start off thinking that what you see is right; end by getting just what you'd expected. And all the while, you remain blissfully unaware that you've done anything other than remain perfectly rational and objective. It's a perfect illusion of validity, and its impact is incredibly difficult to shake, even in circumstances where all logic is against it. (As an example, consider that interviewers tend to make up their minds about a candidate within the first few minutes—and sometimes less—of meeting them. And if the candidate's subsequent behavior paints a different picture, they are still unlikely to alter their opinion—no matter how damning the evidence may be.)

Let's imagine that you need to decide on the suitability of a certain person—let's call her Amy—as a potential teammate. Let me tell you a bit about Amy. First, she is intelligent and industrious.

Stop right there. Chances are you are already thinking, *Okay, yes, great, she would be a wonderful person to work with, intelligent and industrious are both things I'd love to see in a partner.* But what if I was about to continue the statement with, "envious and stubborn"? No longer as good, right? But your initial bias will be remarkably powerful. You will be more likely to discount the latter characteristics and to weigh the former more heavily—all because of your initial intuition. Reverse the two, and the opposite happens; no amount of intelligence and industriousness can save someone who you saw initially as envious and stubborn.

Or consider the following two descriptions of an individual.

> intelligent, skillful, industrious, warm, determined, practical, cautious
> intelligent, skillful, industrious, cold, determined, practical, cautious

If you look at the two lists, you might notice that they are identical, save for one word: *warm* or *cold*. And yet, when study participants heard one of the two descriptions and were then asked to pick which of two traits best described the person (in a list of eighteen pairs from which they always had to choose one trait from each pair), the final impression that the two lists produced was markedly different. Subjects were more likely to find person one generous—and person two the opposite. Yes, you might say, but generosity is an inherent aspect of warmth. Isn't it normal to make that judgment? Let's assume that is the case. Yet participants went a step further in their judgment: they also rated person one in consistently more positive terms than person two, on traits that had nothing whatsoever to do with warmth. Not only did they find person one more sociable and popular (fair enough), but they were also far more likely to think him wise, happy, good natured, humorous, humane, good looking, altruistic, and imaginative.

That's the difference a single word can make: it can color your entire perception of a person, even if every other descriptive point remains the same. And that first impression will last, just as Watson's captivation with Miss Morstan's hair, eyes, and dress will continue to color his evaluation of her as a human being and his perception of what she is and is not capable of doing. We like being consistent and we don't like being wrong. And so, our initial impressions tend to hold an outsized impact, no matter the evidence that may follow.

What about Holmes? Once Mary leaves and Watson exclaims, "What a very attractive woman!" Holmes's response is simple: "Is she? I did not observe." And thereafter follows his admonition to be careful lest personal qualities overtake your judgment.

Does Holmes mean, literally, that he did not observe? Quite the contrary. He observed all of the same physical details as did Watson, and likely far more to boot. What he *didn't* do was make Watson's judgment: that she is a very attractive woman. In that statement, Watson has gone from objective observation to subjective opinion, imbuing physical facts with emotional qualities. That is precisely what Holmes warns against. Holmes may even acknowledge the objective nature of her attractiveness (though if you'll recall, Watson begins by saying that Mary's has "neither regularity of feature nor beauty of complexion"), but he diregards the observation as irrelevant in almost the same breath as he perceives it.

Holmes and Watson don't just differ in the stuff of their attics—in one attic, the furniture acquired by a detective and self-proclaimed loner, who loves music and opera, pipe smoking and indoor target practice, esoteric works on chemistry and renaissance architecture; in the other, that of a war surgeon and self-proclaimed womanizer, who loves a hearty dinner and a pleasant evening out—but in the way their minds organize that furniture to begin with. Holmes knows the biases of his attic like the back of his hand, or the strings of his violin. He knows that if he focuses on a pleasant feeling, he will drop his guard. He knows that if he lets an incidental physical feature get to him, he will run the risk of losing objectivity in the rest of his observation. He knows that if he comes too quickly to a judgment, he will miss much of the evidence against it and pay more attention to the elements that are in its favor. And he knows how strong the pull to act according to a prejudgment will be.

And so he chooses to be selective with those elements that he allows inside his head to begin with. That means with both the furniture that exists already *and* the potential furniture that is vying to get past the hippocampal gateway and make its way into long-term storage. For we should never forget that any experience, any aspect of the world to which we bring our attention is a future memory ready to be made, a new piece of furniture, a new picture to be added to the file, a new element to fit in to our already crowded attics. We can't stop our minds from forming basic judgments. We can't control every piece of information that we retain.

But we can know more about the filters that generally guard our attic's entrance and use our motivation to attend more to the things that matter for our goals—and give less weight to those that don't.

Holmes is not an automaton, as the hurt Watson calls him when he fails to share his enthusiasm for Mary. (He, too, will one day call a woman remarkable—Irene Adler. But only *after* she has bested him in a battle of wits, showing herself to be a more formidable opponent, male or female, than he has ever encountered.) He simply understands that everything is part of a package and could just as well stem from character as from circumstance, irrespective of valence—and he knows that attic space is precious and that we should think carefully about what we add to the boxes that line our minds.

Let's go back to Joe or Jane Stranger. How might the encounter have played out differently had we taken Holmes's approach as a guide? You see Joe's baseball hat or Jane's blue streak, the associations—positive or negative as they may be—come tumbling out. You're feeling like this is the person you do or do not want to spend some time getting to know . . . but before our Stranger opens his mouth, you take just a moment to step back from yourself. Or rather, step more into yourself. Realize that the judgments in your head had to come from somewhere—they always do— and take another look at the person who is making his way toward you. Objectively, is there anything on which to base your sudden impression? Does Joe have a scowl? Did Jane just push someone out of the way? No? Then your dislike is coming from somewhere else. Maybe if you reflect for just a second, you will realize that it is the baseball hat or the blue streak. Maybe you won't. In either case, you will have acknowledged, first off, that you have already predisposed yourself to either like or dislike someone you haven't even met; and second, that you have admitted that you must correct your impression. Who knows, it might have been right. But at least if you reach it a second time, it will be based on objective facts and will come after you've given Joe or Jane a chance to talk. Now you can use the conversation to actually observe—physical details, mannerisms, words. A wealth of evidence that you will treat with the full knowledge that you have already decided, on some level and at some earlier point, to lend more weight to some signs than to others, which you will try to reweigh accordingly.

Maybe Jane is nothing like your friend. Maybe even though you and Joe don't share the same love of baseball, he is actually someone you'd want to get to know. Or maybe you were right all along. The end result isn't as important as whether or not you stopped to recognize that no judgment—no matter how positive or negative, how convincing or seemingly untouchable—begins with an altogether blank slate. Instead, by the time a judgment reaches our awareness, it has already been filtered thoroughly by the interaction of our brain attics and the environment. We can't consciously force ourselves to stop these judgments from forming, but we can learn to understand our attics, their quirks, tendencies, and idiosyncrasies, and to try our best to set the starting point back to a more neutral one, be it in judging a person or observing a situation or making a choice.

A Prime Environment: The Power of the Incidental

In the case of Mary Morstan or Joe and Jane Stranger, elements of physical appearance activated our biases, and these elements were an intrinsic part of the situation. Sometimes, however, our biases are activated by factors that are entirely unrelated to what we are doing—and these elements are sneaky little fellows. Even though they may be completely outside our awareness—in fact, often for that very reason—and wholly irrelevant to whatever it is we're doing, they can easily and profoundly affect our judgment.

At every step, the environment primes us. In the "Adventure of the Copper Beeches," Watson and Holmes are aboard a train to the country. As they pass Aldershot, Watson glances out the window at the passing houses.

> "Are they not fresh and beautiful?" I cried with all the enthusiasm of a man fresh from the fogs of Baker Street.
>
> But Holmes shook his head gravely.
>
> "Do you know, Watson," said he, "that it is one of the curses of a mind with a turn like mine that I must look at everything with

reference to my own special subject. You look at these scattered houses, and you are impressed by their beauty. I look at them, and the only thought which comes to me is a feeling of their isolation and of the impunity with which crime may be committed there."

Holmes and Watson may indeed be looking at the same houses, but what they see is altogether different. Even if Watson manages to acquire all of Holmes's skill in observation, that initial experience will still necessarily differ. For, not only are Watson's memories and habits wholly distinct from Holmes's, but so, too, are the environmental triggers that catch his eye and set his mind thinking along a certain road.

Long before Watson exclaims at the beauty of the passing houses, his mind has been primed by its environment to think in a certain way and to notice certain things. While he is still sitting silently in the train car, he notes the appeal of the scenery, an "ideal spring day, a light blue sky, flecked with fleecy white clouds drifting from west to east." The sun is shining brightly, but there's "an exhilarating nip in the air, which set an edge to a man's energy." And there, in the middle of the new, bright spring leaves, are the houses. Is it all that surprising, then, that Watson sees his world bathed in a pink, happy glow? The pleasantness of his immediate surroundings is priming him to be in a positive mindset.

But that mindset, as it happens, is altogether extraneous in forming other judgments. The houses would remain the same even if Watson were sad and depressed; only his perception of them would likely shift. (Might they not then appear lonely and gloomy?) In this particular case, it little matters whether Watson perceives the houses as friendly or not. But what if, say, he were forming his judgment as a prelude to approaching one of them, be it to ask to use a phone or to conduct a survey or to investigate a crime? Suddenly, how safe the houses are matters a great deal. Do you really want to knock on a door by yourself if there's a chance that the occupants living behind that door are sinister and apt to commit crime with impunity? Your judgment of the house had better be correct—and not the result of a sunny day. Just as we need to know that our internal attics affect our judgment outside of our awareness, so, too, must we be aware of the impact that the external world has on those judgments. Just because something isn't in our attic doesn't mean that it can't influence our attic's filters in very real ways.

There is no such thing as the "objective" environment. There is only our perception of it, a perception that depends in part on habitual ways of thinking (Watson's disposition) and in part on the immediate circumstances (the sunny day). But it's tough for us to realize the extent of the influence that our attic's filters have on our interpretation of the world. When it comes to giving in to the ideal spring day, unprepared Watson is hardly alone—and should hardly be blamed for his reaction. Weather is an extremely powerful prime, one that affects us regularly even though we may have little idea of its impact. On sunny days, to take one example, people report themselves to be happier and to have higher overall life satisfaction than on rainy days. And they have no awareness at all of the connection—they genuinely believe themselves to be more fulfilled as individuals when they see the sun shining in a light blue sky, not unlike the one that Watson sees from his carriage window.

The effect goes beyond simple self-report and plays out in decisions that matter a great deal. On rainy days, students looking at potential colleges pay more attention to academics than they do on sunny days—and for every standard deviation increase in cloud cover on the day of the college visit, a student is 9 percent more likely to actually enroll in that college. When the weather turns gray, financial traders are more likely to make risk-averse decisions; enter the sun, risk-seeking choice increases. The weather does much more than set a pretty scene. It directly impacts what we see, what we focus on, and how we evaluate the world. But do you really want to base a college choice, a judgment of your overall happiness (I'd be curious to see if more divorces or breakups were initiated on rainy days than on sunny days), or a business decision on the state of the atmosphere?

Holmes, on the other hand, is oblivious to the weather—he has been engrossed in his newspaper for the entire train ride. Or rather, he isn't entirely oblivious, but he realizes the importance of focused attention and chooses to ignore the day, much as he had dismissed Mary's attractiveness with an "I haven't noticed." Of course he notices. The question is whether or not he then chooses to attend, to pay attention—and let his attic's contents change in any way as a result. Who knows how the sun would have affected him had he not had a case on his mind and allowed his awareness to wander, but as it is, he focuses on entirely different details and a wholly different context. Unlike Watson, he is understandably anxious and preoccupied. After all, he has just been summoned by a young woman who stated that she had come to her wit's end. He is brooding. He is entirely consumed by the puzzle that he is about to encounter. Is it any surprise then that he sees in the houses a reminder of just the situation that has been preoccupying his mind? It may not be as incidental a prime as the weather has been for Watson, but it is a prime nevertheless.

But, you may (correctly) argue, hasn't Watson been exposed to the exact same telegram by the troubled client? Indeed he has. But for him that matter is far from mind. That's the thing about primes: the way it primes you and the way it primes me may not be the same. Recall the earlier discussion of our internal attic structure, our habitual biases and modes of thought. Those habitual thought patterns have to interact with the environment for the full effect of subtle, preconscious influences on our thought process to take hold; and it is they that largely impact what we notice and how that element then works its way through our minds.

Imagine that I've presented you with sets of five words and have asked you to make four-word sentences out of each set. The words may seem innocuous enough, but hidden among them are the so-called target stimuli: words like *lonely, careful, Florida, helpless, knits,* and *gullible*. Do they remind you of anything? If I lump them all together, they very well might remind you of old age. But spread them out over thirty sets of five-word combinations, and the effect is far less striking—so much less so, in fact, that not a single participant who saw the sentences—of a sample of sixty, in the two original studies of thirty participants each—realized that they had any thematic coherence. But that lack of awareness didn't mean a lack of impact.

If you're like one of the hundreds of people on whom this particular priming task has been used since it was originally introduced in 1996, several things will have happened. You will walk slower now than you did before, and you may even hunch just a little (both evidence of the ideomotor effect of the prime—or its influence on actual physical action).

You'll perform worse on a series of cognitive ability tasks. You'll be slower to respond to certain questions. You may even feel somehow older and wearier than you had previously. Why? You've just been exposed to the Florida effect: a series of age-related stereotypes that, without your awareness, activated a series of nodes and concepts in your brain that in turn prompted you to think and act in a certain fashion. It's priming at its most basic.

Which particular nodes were touched, however, and how the activation spread depends on your own attic and its specific features. If, for instance, you are from a culture that values highly the wisdom of the elderly, while you would have still likely slowed down your walk, you may have become slightly *faster* at the same cognitive tasks. If, on the other hand, you are someone who holds a highly negative attitude toward the elderly, you may have experienced physical effects that were the opposite of those exhibited by the others: you may have walked more quickly and stood up just a bit straighter—to prove that you are unlike the target prime. And that's the point: the prime doesn't exist in a vacuum. Its effects differ. But although individuals may respond differently, they will nevertheless respond.

That, in essence, is why the same telegram may mean something different for Watson and for Holmes. For Holmes, it triggers the expected pattern associated with a mindset that is habitually set to solve crimes. For Watson, it hardly matters and is soon trumped by the pretty sky and the chirping birds. And is that really such a surprise? In general, I think it's safe to suppose that Watson sees the world as a friendlier place than does Holmes. He often expresses genuine amazement at Holmes's suspicions, awe at many of his darker deductions. Where Holmes easily sees sinister intent, Watson notices a beautiful and sympathetic face. Where Holmes brings to bear his encyclopedic knowledge of past crime, and at once applies the past to the present, Watson has no such store to call upon and must rely upon what he does know: medicine, the war, and his brief sojourn with the master detective. Add to that Holmes's tendency, when on an active case and seeking to piece together its details, to drift into the world of his own mind, closing himself off to external distractions that are irrelevant to the subject at hand, as compared to Watson, who is ever happy to note the beauty of a spring day and the appeal of rolling hills, and you have two attics that differ enough in structure and content that they will likely filter just about any input in quite distinct fashion.

We must never forget to factor in the habitual mindset. Every situation is a combination of habitual and in-the-moment goals and motivation—our attic's structure and its current state, so to speak. The prime, be it a sunny day or an anxious telegram or a list of words, may activate our thoughts in a specific direction, but what and how it activates depends on what is inside our attic to begin with and how our attic's structure has been used over time.

But here's the good news: a prime stops being a prime once we're aware of its existence. Those studies of weather and mood? The effect disappeared if subjects were first made explicitly aware of the rainy day: if they were asked about the weather prior to stating their happiness level, the weather no longer had an impact. In studies of the effect of the environment on emotion, if a nonemotional reason is given for a subject's state, the prime effect is likewise eliminated. For instance, in one of the classic studies of emotion, if you're given a shot of adrenaline and then you interact with someone who is displaying strong emotion (which could be either positive or negative), you are likely to mirror that emotion. However, if you are told the shot you received will have physically arousing effects, the mirroring will be mitigated. Indeed, priming studies can be notoriously difficult to replicate: bring any attention at all to the priming mechanism, and you'll likely find the effect go down to zero. When we are aware of the reason for our action, it stops influencing us: we now have something else to which to attribute whatever emotions or thoughts may have been activated, and so, we no longer think that the impetus is coming from our own minds, the result of our own volition.

Activating Our Brain's Passivity

So, how *does* Holmes manage to extricate himself from his attic's instantaneous, pre-attentional judgments? How does he manage to dissociate himself from the external influences that his environment exerts on his mind at any given moment? That very awareness and presence are the key. Holmes has made the passive stage of absorbing information like a leaky sponge—some gets in, some goes in one hole and right out the other, and the sponge has no say or opinion on the process—into an active process, the same type of observation that we will soon discuss in detail. And he has made that active process the brain's default setting.

At the most basic level, he realizes—as now you do—how our thought process begins and why it's so important to pay close attention right from the start. If I were to stop you and explain every reason for your impressions, you may not change them ("But of course I'm still right!"), but at least you will know where they came from. And gradually, you may find yourself catching your mind *before* it leaps to a judgment—in which case you will be far more likely to listen to its wisdom.

Holmes takes nothing, not a single impression, for granted. He does not allow just any trigger that happens to catch his eye to dictate what will or won't make it into his attic and how his attic's contents will or won't be activated. He remains constantly active and constantly vigilant, lest a stray prime worm its way into the walls of his pristine mind space. And while that constant attention may be exhausting, in situations that matter the effort may be well worth it—and with time, we may find that it is becoming less and less effortful.

All it takes, in essence, is to ask yourself the same questions that Holmes poses as a matter of course. *Is something superfluous to the matter at hand influencing my judgment at any given point?* (The answer will almost always be yes.) *If so, how do I adjust my perception accordingly? What has influenced my first impression—and has that first impression in turn influenced others?* It's not that Holmes is not susceptible to priming; it's that he knows its power all too well. So where Watson at once passes judgment on a woman or a country house, Holmes immediately corrects his impression with a *Yes, but.* . . . His message is simple: never forget that an initial impression is only that, and take a moment to reflect on what caused it and what that may signify for your overall aim. Our brains will do certain things as a matter of course, whether or not we want it to. We can't change that. But we can change whether or not we take that initial judgment for granted—or probe it in greater depth. And we should never forget that potent combination of mindfulness and motivation.

In other words, be skeptical of yourself and of your own mind. Observe actively, going beyond the passivity that is our default state. Was something the result of an actual objective behavior (before you term Mary saintly, did you ever observe her doing something that would lead you to believe it?), or just a subjective impression (well, she *looked* so incredibly nice)?

When I was in college, I helped run a global model United Nations conference. Each year we would travel to a different city and invite university students from all over to join in a simulation. My role was committee chair: I prepared topics, ran debates, and, at the end of the conferences, awarded prizes to the students I felt had performed the best. Straightforward enough. Except, that is, when it came to the prizes.

My first year I noticed that Oxford and Cambridge went home with a disproportionate number of speaker awards. Were those students simply that much better, or was there something else going on? I suspected the latter. After all, representatives from the best universities in the world were taking part, and while Oxford and Cambridge were certainly exceptional schools, I didn't know that they would necessarily and consistently have the best delegates. What was going on? Were my fellow award givers somehow, well, biased?

The following year I decided to see if I could find out. I tried to watch my reaction to each student as he spoke, noting my impressions, the arguments that were raised, how convincing the points were, and how persuasively they were argued. And here's where I found something that was rather alarming: to my ear, the Oxford and Cambridge students *sounded* smarter. Put two students next to each other, have them say the exact same thing, and I would like the one with the British accent more. It made no sense whatsoever, but in my mind that accent was clearly activating some sort of stereotype that then biased the rest of the judgment—until, as we neared the end of the conference and the time for prize decisions approached, I was certain that my British delegates were the best of the lot. It was not a pleasant realization.

My next step was to actively resist it. I tried to focus on content alone:

what was each student saying and how was he saying it? Did it add to the discussion? Did it raise points in need of raising? Did it, on the other hand, simply reframe someone else's observation or fail to add anything truly substantive?

I'd be lying if I said the process was easy. Try as I might, I kept finding myself ensnared by the intonation and accent, by the cadence of sentences and not their content. And here it gets truly scary: at the end, I *still* had the urge to give my Oxford delegate the prize for best speaker. She really was the best, I found myself saying. *And aren't I correcting too far in the other direction if I fail to acknowledge as much, in effect penalizing her just for being British? I* wasn't the problem. My awards would be well deserved even if they did happen to go to an Oxford student. It was everyone else who was biased.

Except, my Oxford delegate wasn't the best. When I looked at my painstaking notes, I found several students who had consistently outperformed her. My notes and my memory and impression were at complete odds. In the end, I went with the notes. But it was a struggle up until the last moment. And even after, I couldn't quite kick the nagging feeling that the Oxford girl had been robbed.

Our intuitions are powerful even when entirely inaccurate. And so it is essential to ask, when in the grip of a profound intuition (this is a wonderful person; a beautiful house; a worthy endeavor; a gifted debater): on what is my intuition based? And can I really trust it—or is it just the result of the tricks of my mind? An objective external check, like my committee notes, is helpful, but it's not always possible. Sometimes we just need to realize that even if we are certain we aren't biased in any way, that nothing extraneous is affecting our judgments and choices, chances are that we are not acting in an entirely rational or objective fashion. In that realization—that oftentimes it is best not to trust your own judgment— lies the key to improving your judgment to the point where it can in fact be trusted. What's more, if we are motivated to be accurate, our initial encoding may have less opportunity to spiral out of control to begin with.

But even beyond the realization is the constant practice of the thing. Accurate intuition is really nothing more than practice, of letting skill replace learned heuristics. Just as we aren't inattentive to begin with, we aren't born destined to act in keeping with our faulty thought habits. We just end up doing so because of repeat exposure and practice—and a lack of the same mindful attention that Holmes makes sure to give to his every thought. We may not realize that we have reinforced our brains to think in a certain way, but that is in fact what we have done. And that's both the bad news and the good news—if we taught our brains, we can also unteach them, or teach them differently. Any habit is a habit that can be changed into another habit. Over time, the skill can change the heuristic. As Herbert Simon, one of the founders of what we now call the field of judgment and decision making, puts it, "Intuition is nothing more and nothing less than recognition."

Holmes has thousands of hours of practice on us. His habits have been formed over countless opportunities, twenty-four hours a day, 365 days a year, for every year since his early childhood. It's easy to become discouraged in his presence—but it might, in the end, be more productive to simply become inspired instead. If he can do it, so can we. It will just take time. Habits that have been developed over such an extensive period that they form the very fiber of our minds don't change easily.

Being aware is the first step. Holmes's awareness enables him to avoid many of the faults that plague Watson, the inspectors, his clients, and his adversaries. But how does he go from awareness to something more, something actionable? That process begins with observation: once we understand how our brain attic works and where our thought process originates, we are in a position to direct our attention to the things that matter—and away from the things that don't. And it is to that task of mindful observation that we now turn.

SHERLOCK HOLMES FURTHER READING

"What the deuce is [the solar system] to me?" "I consider that a man's brain originally is like an empty attic . . ." from *A Study in Scarlet*, chapter 2: The Science of Deduction, p. 15.

"Give me problems, give me work . . ." from *The Sign of Four*, chapter 1: The Science of Deduction, p. 5.

"*Miss Morstan entered the room . . .*" "*It is of the first importance not to allow your judgment to be biased by personal qualities.*" from *The Sign of Four,* chapter 2: The Statement of the Case, p. 13.

"*'Are they not fresh and beautiful?' I cried . . .*" from *The Adventures of Sherlock Holmes,* "The Adventure of the Copper Beeches," p. 292.

PART TWO

FROM
OBSERVATION TO
IMAGINATION

CHAPTER THREE

Stocking the Brain Attic:
The Power of Observation

It was Sunday night and time for my dad to whip out the evening's reading. Earlier in the week we had finished *The Count of Monte Cristo*—after a harrowing journey that took several months to complete—and the bar was set high indeed. And there, far from the castles, fortresses, and treasures of France, I found myself face-to-face with a man who could look at a new acquaintance for the first time and proclaim with utter certainty, "You have been in Afghanistan, I perceive." And Watson's reply—"How on earth did you know that?"—was exactly how I immediately felt. How in the world *did* he know that? The matter, it was clear to me, went beyond simple observation of detail.

Or did it? When Watson wonders how Holmes could have possibly known about his wartime service, he posits that someone told the detective beforehand. It's simply impossible that someone could tell such a thing just from . . . looking.

"Nothing of the sort," says Holmes. It is entirely possible. He continues:

I *knew* you came from Afghanistan. From long habit the train of thoughts ran so swiftly through my mind that I arrived at the conclusion without being conscious of intermediate steps. There were such steps, however. The train of reasoning ran, "Here is a gentleman of a medical type, but with the air of a military man. Clearly an army doctor, then. He has just come from the tropics, for his face is dark, and that is not the natural tint of his skin, for his wrists are fair. He has undergone hardship and sickness, as his haggard face says clearly. His left arm has been injured. He holds it in a stiff and unnatural manner. Where in the tropics could an English army doctor have seen much hardship and got his arm wounded?

Clearly in Afghanistan." The whole train of thought did not occupy a second. I then remarked that you came from Afghanistan, and you were astonished.

Sure enough, the starting point seems to be observation, plain and simple. Holmes looks at Watson and gleans at once details of his physical appearance, his demeanor, his manner. And out of those he forms a picture of the man as a whole—just as the real-life Joseph Bell had done in the presence of the astonished Arthur Conan Doyle.

But that's not all. Observation with a capital *O*—the way Holmes uses the word when he gives his new companion a brief history of his life with a single glance—*does* entail more than, well, observation (the lowercase kind). It's not just about the passive process of letting objects enter into your visual field. It is about knowing *what* and *how* to observe and directing your attention accordingly: what details do you focus on? What details do you omit? And how do you take in and capture those details that you do choose to zoom in on? In other words, how do you maximize your brain attic's potential? You don't just throw any old detail up there, if you remember Holmes's early admonitions; you want to keep it as clean as possible. Everything we choose to notice has the potential to become a future furnishing of our attics—and what's more, its addition will mean a change in the attic's landscape that will affect, in turn, each *future* addition. So we have to choose wisely.

Choosing wisely means being selective. It means not only looking but looking properly, looking with real thought. It means looking with the full knowledge that what you note—and how you note it—will form the basis of any future deductions you might make. It's about seeing the full picture, noting the details that matter, and understanding how to contextualize those details within a broader framework of thought.

Why does Holmes note the details he does in Watson's appearance— and why did his real-life counterpart Bell choose to observe what he did in the demeanor of his new patient? ("You see gentlemen," the surgeon told his students, "the man was a respectful man but did not remove his hat. They do not in the army, but he would have learned civilian ways had he been long discharged. He had an air of authority," he continued, "and is obviously Scottish. As to Barbados, his complaint is elephantiasis, which is West Indian and not British, and the Scottish regiments are at present in that particular land." And how did he know which of the many details of the patient's physical appearance were important? That came from sheer practice, over many days and years. Dr. Bell had seen so many patients, heard so many life stories, made so many diagnoses that at some point, it all became natural—just as it did for Holmes. A young, inexperienced Bell would have hardly been capable of the same perspicacity.)

Holmes's explanation is preceded by the two men's discussion of the article "The Book of Life" that Holmes had written for the morning paper—the same article I referred to earlier, which explains how the possibility of an Atlantic or a Niagara could emerge from a single drop of water. After that aqueous start, Holmes proceeds to expand the principle to human interaction.

> Before turning to those moral and mental aspects of the matter which present the greatest difficulties, let the inquirer begin by mastering more elementary problems. Let him, on meeting a fellow-mortal, learn at a glance to distinguish the history of the man, and the trade or profession to which he belongs. Puerile as such an exercise may seem, it sharpens the faculties of observation, and teaches one where to look and what to look for. By a man's finger-nails, by his coat-sleeve, by his boots, by his trouser-knees, by the callosities of his forefinger and thumb, by his expression, by his shirt-cuffs—by each of these things a man's calling is plainly revealed. That all united should fail to enlighten the competent inquirer in any case is almost inconceivable.

Let's consider again how Holmes approaches Watson's stint in Afghanistan. When he lists the elements that allowed him to pinpoint the location of Watson's sojourn, he mentions, in one example of many, a tan in London—something that is clearly *not* representative of that climate and so must have been acquired elsewhere—as illustrating his having arrived from a tropical location. His face, however, is haggard. Clearly then, not a vacation, but something that made him unwell. And his bearing? An unnatural stiffness in one arm, such a stiffness as could result from an injury.

Tropics, sickness, injury: take them together, as pieces of a greater picture, and voilà. Afghanistan. Each observation is taken in context and in tandem with the others—not just as a stand-alone piece but as something that contributes to an integral whole. Holmes doesn't just observe. As he looks, he asks the right questions about those observations, the questions that will allow him to put it all together, to deduce that ocean from the water drop. He need not have known about Afghanistan per se to know that Watson came from a war; he may not have known what to call it then, but he could have well come up with something along the lines of, "You have just come from the war, I perceive." Not as impressive sounding, to be sure, but having the same intent.

As for profession: the category *doctor* precedes *military doctor*—category before subcategory, never the other way around. And about that *doctor*: quite a prosaic guess at a man's profession for someone who spends his life dealing with the spectacular. But prosaic doesn't mean wrong. As you'll note if you read Holmes's other explanations, rarely do his guesses of professions jump—unless with good reason—into the esoteric, sticking instead to more common elements—and ones that are firmly grounded in observation and fact, not based on overheard information or conjecture. A doctor is clearly a much more common profession than, say, a detective, and Holmes would never forget that. Each observation must be integrated into an existing knowledge base. In fact, were Holmes to meet himself, he would categorically *not* guess his own profession. After all, he is the self-acknowledged only "consulting detective" in the world. Base rates—or the frequency of something in a general population—matter when it comes to asking the right questions.

For now, we have Watson, the doctor from Afghanistan. As the good doctor himself says, it's all quite simple once you see the elements that led to the conclusion. But how do we learn to get to that conclusion on our own?

It all comes down to a single word: attention.

Paying Attention Is Anything but Elementary

When Holmes and Watson first meet, Holmes at once correctly deduces Watson's history. But what of Watson's impressions? First, we know he pays little attention to the hospital—where he is heading to meet Holmes for the first time—as he enters it. "It was familiar ground," he tells us, and he needs "no guiding."

When he reaches the lab, there is Holmes himself. Watson's first impression is shock at his strength. Holmes grips his hand "with a strength for which [Watson] should hardly have given him credit." His second is surprise at Holmes's interest in the chemical test that he demonstrates for the newcomers. His third, the first actual observation of Holmes physically: "I noticed that [his hand] was all mottled over with similar pieces of plaster, and discoloured with strong acids." The first two are impressions—or preimpressions—more than observations, much closer to the instinctive, preconscious judgment of Joe Stranger or Mary Morstan in the prior chapter. (Why shouldn't Holmes be strong? It seems that Watson has jumped the gun by assuming him to be somehow akin to a medical student, and thus someone who is not associated with great physical feats. Why shouldn't Holmes be excited? Again, Watson has already imputed his own views of what does and does not qualify as interesting onto his new acquaintance.) The third is an observation in line with Holmes's own remarks on Watson, the observations that lead him to his deduction of service in Afghanistan—except that Watson only makes it because Holmes draws his attention to it by putting a Band-Aid on his finger and remarking on that very fact. "I have to be careful," he explains. "I dabble with poisons a good deal." The only real observation, as it turns out, is one that Watson doesn't actually make until it is pointed out to him.

Why the lack of awareness, the superficial and highly subjective assessment? Watson answers for us when he enumerates his flaws to Holmes—after all, shouldn't prospective flatmates know the worst about each other? "I am extremely lazy," he says. In four words, the essence of the entire problem. As it happens, Watson is far from alone. That fault bedevils most of us—at least when it comes to paying attention. In 1540,

Hans Ladenspelder, a copperplate engraver, finished work on an engraving that was meant to be part of a series of seven: a female, reclining on one elbow on a pillar, her eyes closed, her head resting on her left hand. Peeking out over her right shoulder, a donkey. The engraving's title: "Acedia." The series: The Seven Deadly Sins.

Acedia means, literally, not caring. Sloth. A laziness of the mind that the *Oxford Dictionary* defines as "spiritual or mental sloth; apathy." It's what the Benedictines called the noonday demon, that spirit of lethargy that tempted many a devoted monk to hours of idleness where there should have rightly been spiritual labor. And it's what today might pass for attention deficit disorder, easy distractibility, low blood sugar, or whatever label we choose to put on that nagging inability to focus on what we need to get done.

Whether you think of it as a sin, a temptation, a lazy habit of mind, or a medical condition, the phenomenon begs the same question: why is it so damn hard to pay attention?

It's not necessarily our fault. As neurologist Marcus Raichle learned after decades of looking at the brain, our minds are wired to wander. Wandering is their default. Whenever our thoughts are suspended between specific, discrete, goal-directed activities, the brain reverts to a so-called baseline, "resting" state—but don't let the word fool you, because the brain isn't at rest at all. Instead, it experiences tonic activity in what's now known as the DMN, the default mode network: the posterior cingulate cortex, the adjacent precuneus, and the medial prefrontal cortex. This baseline activation suggests that the brain is constantly gathering information from both the external world and our internal states, and what's more, that it is monitoring that information for signs of something that is worth its attention. And while such a state of readiness could be useful from an evolutionary standpoint, allowing us to detect potential predators, to think abstractly and make future plans, it also signifies something else: our minds are made to wander. *That* is their resting state. Anything more requires an act of conscious will.

The modern emphasis on multitasking plays into our natural tendencies quite well, often in frustrating ways. Every new input, every new demand that we place on our attention is like a possible predator: *Oooh,* says the brain. *Maybe I should pay attention to* that *instead*. And then along comes something else. We can feed our mind wandering ad infinitum. The result? We pay attention to everything and nothing as a matter of course. While our minds might be made to wander, they are not made to switch activities at anything approaching the speed of modern demands. We were supposed to remain ever ready to engage, but not to engage with multiple things at once, or even in rapid succession.

Notice once more how Watson pays attention—or not, as the case may be—when he first meets Holmes. It's not that he doesn't see anything. He notes "countless bottles. Broad, low tables were scattered about, which bristled with retorts, test-tubes, and little Bunsen lamps, with their blue flickering flames." All that detail, but nothing that makes a difference to the task at hand—his choice of future flatmate.

Attention is a limited resource. Paying attention to one thing necessarily comes at the expense of another. Letting your eyes get too taken in by all of the scientific equipment in the laboratory prevents you from noticing anything of significance about the man in that same room. We cannot allocate our attention to multiple things at once and expect it to function at the same level as it would were we to focus on just one activity. Two tasks cannot possibly be in the attentional foreground at the same time. One will inevitably end up being the focus, and the other—or others—more akin to irrelevant noise, something to be filtered out. Or worse still, none will have the focus and all will be, albeit slightly clearer, noise, but degrees of noise all the same.

Think of it this way. I am going to present you with a series of sentences. For each sentence, I want you to do two things: one, tell me if it is plausible or not by writing a *P* for *plausible* or a *N* for *not plausible* by the sentence; and two, memorize the final word of the sentence (at the end of all of the sentences, you will need to state the words in order). You can take no more than five seconds per sentence, which includes reading the sentence, deciding if it's plausible or not, and memorizing the final word. (You can set a timer that beeps at every five-second interval, or find one online—or try to approximate as best you can.) Looking back at a sentence you've already completed is cheating. Imagine that each sentence vanishes once you've read it. Ready?

She was worried about being too hot so she took her new shawl.

She drove along the bumpy road with a view to the sea.

When we add on to our house, we will build a wooden duck.

The workers knew he was not happy when they saw his smile.

The place is such a maze it is hard to find the right hall.

The little girl looked at her toys then played with her doll.

Now please write down the final word of each sentence in order. Again, do not try to cheat by referring back to the sentences.

Done? You've just completed a sentence-verification and span task. How did you do? Fairly well at first, I'm guessing—but it may not have been quite as simple as you'd thought it would be. The mandatory time limit can make it tricky, as can the need to not only read but understand each sentence so that you can verify it: instead of focusing on the last word, you have to process the meaning of the sentence as a whole as well. The more sentences there are, the more complex they become, the trickier it is to tell if they are plausible or not, and the less time I give you per sentence, the less likely you are to be able to keep the words in mind, especially if you don't have enough time to rehearse.

However many words you can manage to recall, I can tell you several things. First, if I were to have you look at each sentence on a computer screen—especially at those times when it was the most difficult for you (i.e., when the sentences were more complex or when you were nearing the end of a list), so that you were keeping more final words in mind at the same time—you would have very likely missed any other letters or images that may have flashed on the screen while you were counting: your eyes would have looked directly at them, and yet your brain would have been so preoccupied with reading, processing, and memorizing in a steady pattern that you would have failed to grasp them entirely. And your brain would have been right to ignore them—it would have distracted you too much to take active note, especially when you were in the middle of your given task.

Consider the policeman in *A Study in Scarlet* who misses the criminal because he's too busy looking at the activity in the house. When Holmes asks him whether the street was empty, Rance (the policeman in question) says, "Well, it was, as far as anybody that could be of any good goes." And yet the criminal was right in front of his eyes. Only, he didn't know how to look. Instead of a suspect, he saw a drunk man—and failed to note any incongruities or coincidences that might have told him otherwise, so busy was he trying to focus on his "real" job of looking at the crime scene.

The phenomenon is often termed attentional blindness, a process whereby a focus on one element in a scene causes other elements to disappear; I myself like to call it attentive inattention. The concept was pioneered by Ulric Neisser, the father of cognitive psychology. Neisser noticed how he could look out a window at twilight and either see the external world or focus on the reflection of the room in the glass. But he couldn't actively pay attention to both. Twilight or reflection had to give. He termed the concept selective looking.

Later, in the laboratory, he observed that individuals who watched two superimposed videos in which people engaged in distinct activities— for instance, in one video they were playing cards, and in the other, basketball—could easily follow the action in either of the films but would miss entirely any surprising event that happened in the other. If, for example, they were watching the basketball game, they would not notice if the cardplayers suddenly stopped playing cards and instead stood up to shake hands. It was just like selective listening—a phenomenon discovered in the 1950s, in which people listening to a conversation with one ear would miss entirely something that was said in their other ear— except, on an apparently much broader scale, since it now applied to multiple senses, not just to a single one. And ever since that initial discovery, it has been demonstrated over and over, with visuals as egregious as people in gorilla suits, clowns on unicycles, and even, in a real-life case, a dead deer in the road escaping altogether the notice of people who were staring directly at them.

Scary, isn't it? It should be. We are capable of wiping out entire chunks of our visual field without knowingly doing so. Holmes admonished Watson for seeing but not observing. He could have gone a step further: sometimes we don't even see.

We don't even need to be actively engaged in a cognitively demanding task to let the world pass us by without so much as a realization of what we're missing. For instance, when we are in a foul mood, we quite literally see less than when we are happy. Our visual cortex actually takes in less information from the outside world. We could look at the exact same scene twice, once on a day that has been going well and once on a day that hasn't, and we would notice less—and our brains would take in less—on the gloomy day.

We can't actually be aware unless we pay attention. No exceptions. Yes, awareness may require only minimal attention, but it does require some attention. Nothing happens quite automatically. We can't be aware of something if we don't attend to it.

Let's go back to the sentence-verification task for a moment. Not only will you have missed the proverbial twilight for focusing too intently at the reflection in the window, but the harder you were thinking, the more dilated your pupils will have become. I could probably tell your mental effort—as well as your memory load, your ease with the task, your rate of calculation, and even the neural activity of your locus coeruleus (the only source in the brain of the neurotransmitter norepinephrine and an area implicated in memory retrieval, a variety of anxiety syndromes, and selective attentional processing), which will also tell me whether you are likely to keep going or to give up—just by looking at the size of your pupils.

But there is one encouraging thing: the importance—and effectiveness—of training, of brute practice, is overwhelmingly clear. If you were to do the sentence verification regularly—as some subjects did in fact do—your pupils would gradually get smaller; your recall would get more natural; and, miracle of all miracles, you'd notice those same letters or images or whatnot that you'd missed before. You'd probably even ask yourself, how in the world did I *not* see this earlier? What was previously taxing will have become more natural, more habitual, more effortless; in other words, easier. What used to be the purview of the Holmes system would have sneaked into the Watson system. And all it will have taken is a little bit of practice, a small dose of habit formation. Your brain can be one quick study if it wants to be.

The trick is to duplicate that same process, to let your brain study and learn and make effortless what was once effortful, in something that lacks the discrete nature of a cognitive task like the sentence verification, in something that is so basic that we do it constantly, without giving it much thought or attention: the task of looking and thinking.

Daniel Kahneman argues repeatedly that System 1—our Watson system—is hard to train. It likes what it likes, it trusts what it trusts, and that's that. His solution? Make System 2—Holmes—do the work by taking System 1 forcibly out of the equation. For instance, use a checklist of characteristics when hiring a candidate for a job instead of relying on your impression, an impression that, as you'll recall, is formed within the first five minutes or less of meeting someone. Write a checklist of steps to follow when making a diagnosis of a problem, be it a sick patient, a broken car, writer's block, or whatever it is you face in your daily life, instead of trying to do it by so-called instinct. Checklists, formulas, structured procedures: those are your best bet—at least, according to Kahneman.

The Holmes solution? Habit, habit, habit. That, and motivation. Become an expert of sorts at those types of decisions or observation that you want to excel at making. Reading people's professions, following their trains of thought, inferring their emotions and thinking from their demeanor? Fine. But just as fine are things that go beyond the detective's purview, like learning to tell the quality of food from a glance or the proper chess move from a board or your opponent's intention in baseball, poker, or a business meeting from a gesture. If you learn first how to be selective accurately, in order to accomplish precisely what it is you want to accomplish, you will be able to limit the damage that System Watson can do by preemptively teaching it to not muck it up. The important thing is the proper, selective training—the presence of mind—coupled with the desire and the motivation to master your thought process.

No one says it's easy. When it comes right down to it, there is no such thing as free attention; it all has to come from somewhere. And every time we place an additional demand on our attentional resources—be it by listening to music while walking, checking our email while working, or following five media streams at once—we limit the awareness that surrounds any one aspect and our ability to deal with it in an engaged, mindful, and productive manner.

What's more, we wear ourselves out. Not only is attention limited, but it is a finite resource. We can drain it down only so much before it needs a reboot. Psychologist Roy Baumeister uses the analogy of a muscle to talk about self-control—an analogy that is just as appropriate when it comes to attention: just as a muscle, our capacity for self-control has only so many exertions in it and will get tired with too much use. You need to replenish a muscle—actually, physically replenish it, with glucose and a rest period; Baumeister is not talking about metaphorical energy—though a psych-you-up speech never hurt—to remain in peak form. Otherwise, performance will flag. Yes, the muscle will get bigger with use (you'll improve your self-control or attentional ability and be able to exercise it for longer and longer periods and at more complicated tasks), but its growth, too, is limited. Unless you take steroids—the exercise equivalent of a Ritalin or Adderall for superhuman attention—you will reach your limit, and even steroids take you only so far. And failure to use it? It will shrink right back to its pre-exercise size.

Improving Our Natural Attentional Abilities

Picture this. Sherlock Holmes and Dr. Watson are visiting New York (not so far-fetched—their creator spent some memorable time in the city) and decide to go to the top of the Empire State Building. When they arrive at the observation deck, they are accosted by a quirky stranger who proposes a contest: which of them will be the first to spot an airplane in flight? They can use any of the viewing machines—in fact, the stranger even gives them each a stack of quarters—and look wherever they'd like. The only consideration is who sees the plane first. How do the two go about the task?

It may seem like an easy thing to do: an airplane is a pretty large bird, and the Empire State Building is a pretty tall house, with a pretty commanding 360-degree view. But if you want to be first, it's not as simple as standing still and looking up (or over). What if the plane is somewhere else? What if you can't see it from where you're standing? What if it's behind you? What if you could have been the first to spot one that was farther away if only you'd used your quarters on a viewing machine instead of standing there like an idiot with only your naked eyes? There are a lot of what-ifs—if you want to emerge victorious, that is—but they can be made manageable what-ifs, if you view them as nothing more than a few strategy choices.

Let's first imagine how Watson would go about the task. Watson, as we know, is energetic. He is quick to act and quick to move. And he's also quite competitive with Holmes—more than once, he tries to show that he, too, can play the detective game; there's nothing he likes more than thinking he can beat Holmes on his own turf. I'm willing to bet that he'll do something like the following. He won't waste a single moment in thought (Time's a ticking! Better move quickly). He'll try to cover as much ground as possible (It could come from anywhere! And I don't want to be the idiot who's left behind, that's for sure!) and will thus likely plop coins into as many machines as he can find and then run between them, scanning the horizon in between sprints. He may even experience a few false alarms (It's a plane! Oh, no, it's a bird) in his desire to spot something—and when he does, he'll genuinely think that he's seeing a plane. And in between the running and the false spotting, he'll quickly run out of breath. *This is horrid*, he'll think. *I'm exhausted.* And anyway, what's the point? It's a stupid airplane. Let's hope for his sake that a real plane comes quickly.

What of Holmes? I propose that he'd first orient himself, doing some quick calculation on the location of the airports and thus the most likely direction of a plane. He would even, perhaps, factor in such elements as the relative likelihood of seeing a plane that's taking off or landing given the time of day and the likeliest approach or takeoff paths, depending on the answer to the former consideration. He would then position himself so as to focus in on the area of greatest probability, perhaps throwing a coin in a machine for good measure and giving it a quick once-over to make sure he isn't missing anything. He would know when a bird was just a bird, or a passing shadow just a low-hanging cloud. He wouldn't rush. He would look, and he would even listen, to see if a telltale noise might help direct his attention to a looming jet. He might even smell and feel the air for changing wind or a whiff of gasoline. All the while, he'd be rubbing together his famous long-fingered hands, thinking, *Soon; it will come soon. And I know precisely where it will appear.*

Who would win? There's an element of chance involved, of course, and either man could get lucky. But play the game enough times, and I'd be willing to bet that Holmes would come out on top. While his strategy may at first glance seem slower, not nearly as decisive, and certainly not as inclusive as Watson's, at the end it would prove to be the superior of the two.

Our brains aren't stupid. Just as we remain remarkably efficient and effective for a remarkable percentage of the time despite our cognitive biases, so, too, our Watsonian attentional abilities are as they are for a reason. We don't notice everything because noticing everything—each sound, each smell, each sight, each touch—would make us crazy (in fact, a lack of filtering ability is the hallmark of many psychiatric conditions). And Watson had a point back there: searching for that airplane? Perhaps not the best use of his time.

You see, the problem isn't a lack of attention so much as a lack of mindfulness and direction. In the usual course of things, our brains pick and choose where to focus without much conscious forethought on our part. What we need to learn instead is how to tell our brains what and how to filter, instead of letting them be lazy and decide for us, based on what they think would make for the path of least resistance.

Standing on top of the Empire State Building, watching quietly for airplanes, Sherlock Holmes has illustrated the four elements most likely to allow us to do just that: selectivity, objectivity, inclusivity, and engagement.

1. Be Selective

Picture the following scene. A man passes by a bakery on his way to the office. The sweet smell of cinnamon follows him down the street. He pauses. He hesitates. He looks in the window. The beautiful glaze. The warm, buttery rolls. The rosy doughnuts, kissed with a touch of sugar. He goes in. He asks for a cinnamon roll. *I'll go on my diet tomorrow,* he says. *You only live once. And besides, today is an exception. It's brutally cold and I have a tough meeting in just an hour.*

Now rewind and replay. A man passes by a bakery on his way to the office. He smells cinnamon. *I don't much care for cinnamon, now that I think about it,* he says. *I far prefer nutmeg, and there isn't any here that I can smell.* He pauses. He hesitates. He looks in the window. The oily, sugary glaze that has likely caused more heart attacks and blocked arteries than you can count. The dripping rolls, drenched in butter—actually, it's probably margarine, and everyone knows you can't make good rolls with that. The burned doughnuts that will sit like lumps in your stomach and make you wonder why you ever ate them to begin with. *Just as I thought,* he says. *Nothing here for me.* He walks on, hurrying to his morning meeting. *Maybe I'll have time to get coffee before,* he thinks.

What has changed between scenario one and scenario two? Nothing visible. The sensory information has remained identical. But somehow our hypothetical man's mindset has shifted—and that shift has, quite literally, affected how he experiences reality. It has changed how he is processing information, what he is paying attention to, and how his surroundings interact with his mind.

It's entirely possible. Our vision is highly selective as is—the retina normally captures about ten billion bits per second of visual information, but only ten thousand bits actually make it to the first layer of the visual cortex, and, to top it off, only 10 percent of the area's synapses is dedicated to incoming visual information at all. Or, to put it differently, our brains are bombarded by something like eleven million pieces of data—that is, items in our surroundings that come at all of our senses—at once. Of that, we are able to consciously process only about forty. What that basically means is that we "see" precious little of what's around us, and what we think of as objective seeing would better be termed selective filtering—and our state of mind, our mood, our thoughts at any given moment, our motivation, and our goals can make it even more picky than it normally is.

It's the essence of the cocktail party effect, when we note our name out of the din of a room. Or of our tendency to notice the very things we are thinking about or have just learned at any given moment: pregnant women noticing other pregnant women everywhere; people noting the dreams that then seem to come true (and forgetting all of the others); seeing the number 11 everywhere after 9/11. Nothing in the environment actually changes—there aren't suddenly more pregnant women or prescient dreams or instances of a particular number—only your state does. That's why we are so prone to the feeling of coincidence: we forget all those times we were wrong or nothing happened and remember only the moments that matched—because those are the ones we paid attention to in the first place. As one Wall Street guru cynically observed, the key to being seen as a visionary is to always make your predictions in opposing pairs. People will remember those that came true and promptly forget those that didn't.

Our minds are set the way they are for a reason. It's exhausting to have the Holmes system running on full all the time—and not very productive at that. There's a reason we're prone to filter out so much of our environment: to the brain, it's noise. If we tried to take it all in, we wouldn't last very long. Remember what Holmes said about your brain attic? It's precious real estate. Tread carefully and use it wisely. In other words, be selective about your attention.

At first glance, this may seem counterintuitive: after all, aren't we trying to pay attention to more, not less? Yes, but the crucial distinction is between quantity and quality. We want to learn to pay attention *better*, to become superior observers, but we can't hope to achieve this if we thoughtlessly pay attention to everything. That's self-defeating. What we need to do is allocate our attention mindfully. And mindset is the beginning of that selectivity.

Holmes knows this better than anyone. True, he can note in an instant the details of Watson's attire and demeanor, the furnishings of a room down to the most minute element. But he is just as likely to not notice the weather outside or the fact that Watson has had time to leave the apartment and return to it. It is not uncommon for Watson to point out that a storm is raging outside, only to have Holmes look up and say that he hadn't noticed—and in *Sherlock*, you will often find Holmes speaking to a blank wall long after Watson has retired, or left the apartment altogether.

Whatever the situation, answering the question of what, specifically, you want to accomplish will put you well on your way to knowing how to maximize your limited attentional resources. It will help direct your mind, prime it, so to speak, with the goals and thoughts that are actually important—and help put those that aren't into the background. Does your brain notice the sweet smell or the grease on the napkin? Does it focus on Watson's tan or the weather outside?

Holmes doesn't theorize before he has the data, it's true. But he does form a precise plan of attack: he defines his objectives and the necessary elements for achieving them. So in *The Hound of the Baskervilles*, when Dr. Mortimer enters the sitting room, Holmes already knows what he wants to gain from the situation. His last words to Watson before the gentleman's entrance are, "What does Dr. James Mortimer, the man of science, ask of Sherlock Holmes, the specialist in crime?" Holmes hasn't yet met the man in question, but already he knows what his observational goal will be. He has defined the situation before it even began (and has managed to examine the doctor's walking stick to boot).

When the doctor does appear, Holmes sets at once to ascertain the purpose of his visit, asking about every detail of the potential case, the people involved, the circumstances. He learns the history of the Baskerville legend, the Baskerville house, the Baskerville family. He inquires to the neighbors, the occupants of the Baskerville estate, the doctor himself, insofar as he relates to the family. He even sends for a map of the area, so that he can gather the full range of elements, even those that may have been omitted in the interview. Absolute attention to every element that bears on his original goal: to solve that which Dr. James Mortimer asks of Sherlock Holmes.

As to the rest of the world in between the doctor's visit and the evening, it has ceased to exist. As Holmes tells Watson at the end of the day, "My body has remained in this armchair and has, I regret to observe, consumed in my absence two large pots of coffee and an incredible amount of tobacco. After you left I sent down to Stamford's for the Ordnance map of this portion of the moor, and my spirit has hovered over it all day. I flatter myself that I could find my way about."

Holmes has visited Devonshire in spirit. What his body did, he does not know. He isn't even being entirely facetious. Chances are he really wasn't aware of what he was drinking or smoking—or even that the air in the room has become so unbreathable that Watson is forced to open all of the windows the moment he returns. Even Watson's excursion into the outside world is part of Holmes's attentional plan: he expressly asks his flatmate to leave the apartment so as not to distract him with needless inputs.

So, noticing everything? Far from it, despite the popular conception of the detective's abilities. But noticing everything that matters to the purpose at hand. And therein lies the key difference. (As Holmes notes in "Silver Blaze" when he finds a piece of evidence that the inspector had overlooked, "I only saw it because I was looking for it." Had he not had an a priori reason for the search, he never would have noticed it—and it wouldn't have really mattered, not for him, at least.) Holmes doesn't waste his time on just anything. He allocates his attention strategically.

So, too, we must determine our objective in order to know what we're looking for—and where we're looking for it. We already do this naturally in situations where our brains know, without our having to tell them, that something is important. Remember that party in chapter two, the one with that girl with the blue streak in her hair and that guy whose name you can't be bothered to remember? Well, picture yourself back in that group, chatting away. Look around and you'll notice many groups just like yours, spread all around the room. And just like yours, they are all chatting away. Talk, talk, talk, talk, talk. It's exhausting if you stop to think about it, all this talking going on nonstop. That's why you ignore it. It becomes background noise. Your brain knows how to take the environment and tune out most of it, according to your general goals and needs (specifically, dorsal and ventral regions in the parietal and frontal cortex become involved in both goal-directed—parietal—and stimulus-driven—frontal—attentional control). At the party, it's focusing on the conversation you are having and treating the rest of the words—some of which may be at the exact same volume—as meaningless chatter.

And all of a sudden one conversation comes into clear focus. It's not chatter anymore. You can hear every word. You turn your head. You snap

to attention. What just happened? Someone said your name, or something that sounded like your name. That was enough to signal to your
brain to perk up and focus. Here was something that had relevance to
you; pay attention. It's what's known as the classic cocktail party effect:
one mention of your name, and neural systems that were sailing along
snap into action. You don't even have to do any work.

Most things don't have such nicely built-in flags to alert you to their
significance. You need to teach your mind to perk up, as if it were hearing
your name, but absent that oh-so-clear stimulus. You need, in Holmes's
words, to know what you are looking for in order to see it. In the case of
the man walking past the bakery, it's simple enough. Discrete goal: not to
eat the baked goods. Discrete elements to focus on: the sweets themselves
(find the negative in their appearance), the smells (why not focus on the
exhaust smell from the street instead of the sweet baking? or burnt coffee?), and the overall environment (think forward to the meeting, to the
wedding and the tuxedo, instead of zoning in on the current stimuli). I'm
not saying that it's actually easy to do—but at least the top-down processing that needs to happen is clear.

But what about making a decision, solving a problem at work, or
something even more amorphous? It works the same way. When psychologist Peter Gollwitzer tried to determine how to enable people to set
goals and engage in goal-directed behavior as effectively as possible, he
found that several things helped improve focus and performance: (1)
thinking ahead, or viewing the situation as just one moment on a larger,
longer timeline and being able to identify it as just one point to get past in
order to reach a better future point; (2) being specific and setting specific
goals, or defining your end point as discretely as possible and pooling
your attentional resources as specifically as you can; (3) setting up if/
then contingencies, or thinking through a situation and understanding
what you will do if certain features arise (i.e., if I catch my mind wandering, then I will close my eyes, count to ten, and refocus); (4) writing everything down instead of just thinking it in your head, so that you maximize
your potential and know in advance that you won't have to try to re-create
anything from scratch; and (5) thinking of both repercussions—what

would happen should you fail—and of positive angles, the rewards if you succeed.

Selectivity—mindful, thoughtful, smart selectivity—is the key first step to learning how to pay attention and make the most of your limited resources. Start small; start manageable; start focused. System Watson may take years to become more like System Holmes, and even then it may never get there completely, but by being mindfully focused, it can sure get closer. Help out the Watson system by giving it some of the Holmes system's tools. On it's own, it's got nothing.

One caveat, however: you can set goals to help you filter the world, but be careful lest you use these goals as blinders. Your goals, your priorities, your answer to the "what I want to accomplish" question must be flexible enough to adapt to changing circumstances. If the available information changes, so should you. Don't be afraid to deviate from a preset plan when it serves the greater objective. That, too, is part of the observational process.

Let your inner Holmes show your inner Watson where to look. And don't be like Inspector Alec MacDonald, or Mac, as Holmes calls him. Listen to what Holmes suggests, be it a change of course or a walk outside when you'd rather not.

2. Be Objective

In "The Adventure of the Priory School," a valuable pupil goes missing from a boarding school. Also vanished is the school's German master. How could such a calamity occur in a place of such honor and prestige, termed "without exception, the best and most select preparatory school in England"? Dr. Thorneycroft Huxtable, the school's founder and principal, is flummoxed in the extreme. By the time he makes it from the north of England to London, to consult with Mr. Holmes, he is so overwrought that he proceeds at once to collapse, "prostrate and insensible," upon the bearskin hearth rug of 221B Baker Street.

Not one but two people missing—and the pupil, the son of the Duke of Holdernesse, a former cabinet minister and one of the wealthiest men in England. It must certainly be the case, Huxtable tells Holmes, that Heidegger, the German master, was somehow an accomplice to the disappearance. His bicycle is missing from the bicycle shed and his room bears signs of a hasty exit. A kidnapper? A kidnapper's accomplice? Huxtable can't be sure, but the man can hardly be blameless. It would be too much to chalk the double disappearance off to something as simple as coincidence.

A police investigation is initiated at once, and when a young man and boy are seen together on an early train at a neighboring station, it seems that the policemen have done their duty admirably. The investigation is duly called off. Quite to Huxtable's chagrin, however, it soon becomes clear that the couple in question is altogether unrelated to the disappearance. And so, three days after the mysterious events, the principal has come to consult Mr. Holmes. Not a moment too soon, says the detective— and perhaps, several moments too late. Precious time has been lost. Will the fugitives be found before even greater tragedy occurs?

What makes up a situation like this? Answering that question is not as easy as stating a series of facts—missing boy, missing instructor, missing bike, and the like—or even delineating each one of the accompanying details—state of the boy's room, state of the instructor's room, clothing, windows, plants, etcetera. It also entails understanding something very specific: a situation (in its broadest sense, be it mental, physical, or something as un-situation-like as an empty room) is inherently dynamic. And you, by the very action of entering into it, shift it from what it was before your arrival to something altogether different.

It's Heisenberg's uncertainty principle in action: the fact of observing changes the thing being observed. Even an empty room is no longer the same once you're inside. You cannot proceed as if it hadn't changed. This may sound like common sense, but it is actually much harder to understand in practice than it seems in theory.

Take, for instance, a commonly studied phenomenon known as the white coat effect. Maybe you have an ache or a cough that you want to check out. Maybe you are simply overdue for your next physical. You sigh, pick up the phone, and make an appointment with your doctor. The next day you make your way to his office. You sit in the waiting room. Your name is called. You go in for your appointment.

It's safe to assume that the you that is walking in to get the checkup is the same you that placed the call, right? Wrong. Study after study has shown that for many people, the mere fact of entering a doctor's office and seeing the physician—hence, the white coat—is enough to significantly alter vital signs. Pulse, blood pressure, even reactions and blood work can all change simply because you are seeing a doctor. You may not even feel particularly anxious or stressed. All the same, your readings and results will have changed. The situation has shifted through mere presence and observation.

Recall Dr. Huxtable's view of the events surrounding the disappearance: there is a fugitive (the boy), an accomplice (the tutor), and a bike stolen for purposes of flight or deceit. Nothing more, nothing less. What the principal reports to Holmes is fact (or so he believes).

But is it really? It's psychologist Daniel Gilbert's theory about believing what we see taken a step further: we believe what we *want* to see and what our mind attic decides to see, encode that belief instead of the facts in our brains, and then think that we saw an objective fact when really what we remember seeing is only our limited perception at the time. We forget to separate the factual situation from our subjective interpretation of it. (One need only look at the inaccuracy of expert witness testimonies to see how bad we are at assessing and remembering.) Because the school's principal at once suspected a kidnapping, he has noticed and reported the very details that support his initial idea—and hasn't taken the time to get the full story in the least. And yet, he has no clue that he is doing it. As far as he's concerned, he remains entirely objective. As the philosopher Francis Bacon put it, "The human understanding when it has once adopted an opinion (either as being the received opinion or as being agreeable to itself) draws all things else to support and agree with it." True objectivity can never be achieved—even the scientific objectivity of Holmes isn't ever complete—but we need to understand just how far we stray in order to approximate a holistic view of any given situation.

Setting your goals beforehand will help you direct your precious attentional resources properly. It should *not* be an excuse to reinterpret objective facts to mesh with what you want or expect to see. Observation and deduction are two separate, distinct steps—in fact, they don't even come one right after the other. Think back for a moment to Watson's Af-ghanistan sojourn. Holmes stuck to objective, tangible facts in his obser-vations. There was no extrapolation at first; that happened only after. And he always asked how those facts could fit together. Understanding a situation in its fullness requires several steps, but the first and most fun-damental is to realize that observation and deduction are not the same. To remain as objective as you possibly can.

My mother was quite young—unbelievably young, by today's stan-dards; average by those of 1970s Russia—when she gave birth to my older sister. My sister was quite young when she gave birth to my niece. I can-not even begin to list the number of times that people—from complete strangers to mothers of classmates and even waiters in restaurants—have thought they were seeing one thing and acted according to that thought, when in reality they were seeing something entirely different. My mother has been taken for my sister's sister. These days, she is routinely taken for my niece's mother. Not grave errors on the observer's part, to be sure, but errors nonetheless—and errors which have, in many cases, gone on to affect both their behavior and their subsequent judgments and reactions. It's not just a question of mixing up generations. It's also a question of applying modern American values to the behavior of women in Soviet Russia—an entirely different world. In American lingo, Mom was a teen-age mother. In Russia, she was married and not even the first among her friends to have a child. It was just the way things were done.

You think; you judge; and you don't think twice about what you've just done.

Hardly ever, in describing a person, an object, a scene, a situation, an interaction do we see it as just a valueless, objective entity. And hardly ever do we consider the distinction—since, of course, it hardly ever mat-ters. But it's the rare mind that has trained itself to separate the objective fact from the immediate, subconscious, and automatic subjective inter-pretation that follows.

The first thing Holmes does when he enters a scene is to gain a sense of what has been going on. Who has touched what, what has come from where, what is there that shouldn't be, and what isn't there that should be. He remains capable of extreme objectivity even in the face of extreme circumstances. He remembers his goal, but he uses it to filter and not to inform. Watson, on the other hand, is not so careful.

Consider again the missing boy and the German schoolmaster. Unlike Dr. Huxtable, Holmes understands that a situation is colored by his interpretation. And so, unlike the headmaster, he entertains the possibility that the so-called facts are not what they seem. The principal is severely limited in his search by one crucial detail: he—along with everyone else—is looking for a fugitive and an accomplice. But what if Herr Heidegger is nothing of the sort? What if he isn't fleeing but doing something else entirely? The missing boy's father supposes he might be helping the lad flee to his mother in France. The principal, that he might be conducting him to another location. The police, that they have escaped on a train. But not a single person save Holmes realizes that the story is merely that. They are not to look for a fleeing schoolmaster, wherever the destination may be, but for the schoolmaster (no modifier necessary) and the boy, and not necessarily in the same place. Everyone interprets the missing man as somehow involved in the disappearance, be it as accomplice or instigator. No one stops to consider that the only available evidence points to nothing beside the fact that he's missing.

No one, that is, except for Sherlock Holmes. He realizes that he is looking for a missing boy. He is also looking for a missing schoolmaster. That is all. He lets any additional facts emerge as and when they may. In this more evenhanded approach, he chances upon a fact that has completely passed by the school director *and* the police: that the schoolmaster hasn't fled with the boy at all and is instead lying dead nearby, "a tall man, full-bearded, with spectacles, one glass of which had been knocked out. The cause of his death was a frightful blow upon the head, which had crushed in part of his skull."

To find the body, Holmes doesn't discover any new clues; he just knows to look at what is there in an objective light, without preconception or preformed theories. He enumerates the steps that led to his discovery to Watson:

"Let us continue our reconstruction. He meets his death five miles from the school—not by a bullet, mark you, which even a lad might conceivably discharge, but by a savage blow dealt by a vigorous arm. The lad, then, *had* a companion in his flight. And the flight was a swift one, since it took five miles before an expert cyclist could overtake them. Yet we surveyed the ground round the scene of the tragedy. What do we find? A few cattle tracks, nothing more. I took a wide sweep round, and there is no path within fifty yards. Another cyclist could have had nothing to do with the actual murder, nor were there any human foot-marks."

"Holmes," I cried, "this is impossible."

"Admirable!" he said. "A most illuminating remark. It *is* impossible as I state it, and therefore I must in some respect have stated it wrong. Yet you saw for yourself. Can you suggest any fallacy?"

Watson cannot. Instead, he suggests that they give up altogether. "I am at my wit's end," he says.

"Tut, tut," scolds Holmes. "We have solved some worse problems. At least we have plenty of material, if we can only use it."

In this brief exchange, Holmes has shown that all of the headmaster's theories were misguided. There were at least three people, not at most two. The German instructor was trying to save the boy, not hurt him or flee with him (the most likely scenario, given his now-dead state and the fact that he followed the initial tire tracks and had to overtake the fleeing boy; clearly, he could be neither kidnapper nor accomplice). The bike was a means of pursuit, not stolen property for some sinister motive. And what's more, there must have been another bike present to aid the escape of the boy and unidentified other or others. Holmes hasn't done anything spectacular; he has just allowed the evidence to speak. And he has followed it without allowing himself to skew the facts to conform with the situation. In short, he has behaved with the coolness and reflection of System Holmes, while Huxtable's conclusions show every marking of the hot, reflexive, leap-before-you-look school of System Watson.

To observe, you must learn to separate situation from interpretation, yourself from what you're seeing. System Watson wants to run away into the world of the subjective, the hypothetical, the deductive. Into the world that would make the most sense to you. System Holmes knows to hold back the reins.

A helpful exercise is to describe the situation from the beginning, either out loud or in writing, as if to a stranger who isn't aware of any of the specifics—much like Holmes talks his theories through out loud to Watson. When Holmes states his observations in this way, gaps and inconsistencies that weren't apparent before come to the surface.

It's an exercise not unlike reading your own work out loud to catch any errors in grammar, logic, or style. Just like your observations are so entwined with your thoughts and perception that you may find it difficult, if not impossible, to disentangle the objective reality from its subjective materialization in your mind, when you work on an essay or a story or a paper, or anything else really, you become so intimately acquainted with your own writing that you are liable to skip over mistakes and to read what the words should say instead of what they do say. The act of speaking forces you to slow down and catch those errors that are invisible to your eyes. Your ear notes them when your eye does not. And while it may seem a waste of time and effort to reread mindfully and attentively, out loud, it hardly ever fails to yield a mistake or flaw that you would have otherwise missed.

It's easy to succumb to Watson's conflating logic, to Huxtable's certainty in what he says. But every time you find yourself making a judgment immediately upon observing—in fact, even if you don't think you are, and even if everything seems to make perfect sense—train yourself to stop and repeat: *It is impossible as I state it, and therefore I must in some respect have stated it wrong.* Then go back and restate it from the beginning and in a different fashion than you did the first time around. Out loud instead of silently. In writing instead of in your head. It will save you from many errors in perception.

3. Be Inclusive

Let's go back for a moment to *The Hound of the Baskervilles.* In the early chapters of the story, Henry Baskerville, the heir to the Baskerville estate, reports that his boot has gone missing. But not just one boot. Henry finds that the missing boot has miraculously reappeared the day after its disappearance—only to discover that a boot from another pair has vanished in its stead. To Henry this is annoying but nothing more. To Sherlock Holmes it is a key element in a case that threatens to devolve into a paranormal, voodoo-theory-generating free-for-all. What to others is a mere curiosity to Holmes is one of the more instructive points in the case: the "hound" they are dealing with is an actual animal, not a phantasm. An animal who relies on his sense of smell in a fundamental fashion. As Holmes later tells Watson, the exchange of one stolen boot for another was "a most instructive incident, since it proved conclusively to my mind that we were dealing with a real hound, as no other supposition could explain this anxiety to obtain an old boot and this indifference to a new one."

But that's not all. Apart from the vanishing boot, there is the issue of a more obvious warning. While consulting with Holmes in London, Henry has received anonymous notes that urge him to stay away from Baskerville Hall. Once again, to everyone but Holmes these notes are nothing more than what they seem. For Holmes they form the second part of the key to the case. As he tells Watson:

> "It may possibly recur to your memory that when I examined the paper upon which the printed words were fastened I made a close inspection for the water-mark. In doing so I held it within a few inches of my eyes, and was conscious of a faint smell of the scent known as white jessamine. There are seventy-five perfumes, which it is very necessary that a criminal expert should be able to distinguish from each other, and cases have more than once within my own experience depended upon their prompt recognition. The scent suggested the presence of a lady, and already my thoughts began to turn toward the Stapletons. Thus I had made certain of the hound, and had guessed at the criminal before we ever went to the west country."

There it is a second time: smell. Holmes doesn't just read the note and look at it. He also smells it. And in the scent, not in the words or the appearance, is where he finds the clue that helps him identify the possible criminal. Absent smell, two central clues of the case would remain unidentified—and so they do to everyone but the detective. I am not suggesting you go out and memorize seventy-five perfumes. But you should never neglect your sense of smell—or indeed any of your other senses—because they certainly won't neglect you.

Consider a scenario where you're buying a car. You go to the dealer and look at all the shining specimens sitting out on the lot. How do you decide which model is the right one for you? If I ask you that question right now, you will likely tell me that you'd weigh any number of factors, from cost to safety, appearance to comfort, mileage to gas use. Then you'll pick the vehicle that best matches your criteria.

But the reality of the situation is far more complex. Imagine, for instance, that at the moment you're in the lot, a man walks by with a mug of steaming hot chocolate. You might not even remember that he passed, but the smell triggers memories of your grandfather: he used to make you hot chocolate when you spent time together. It was your little ritual. And before you know it, you're leaving the lot with a car like the one your grandfather drove—and have conveniently forgotten (or altogether failed to note) its less-than-stellar safety rating. And you very likely don't even know why exactly you made the choice you did. You're not wrong per se, but your selective remembering might mean a choice that you'll later regret.

Now imagine a different scenario. This time there's a pervasive smell of gasoline: the lot is across the street from a gas station. And you remember your mother warning you to be careful around gas, that it could catch fire, that you could get hurt. Now you're focused on safety. You'll likely be leaving the lot with a car that is quite different from your grandfather's. And again, you may not know why.

Up to now, I've been talking about attention as a visual phenomenon. And it is, for the most part. But it is also much more. Remember how in the hypothetical foray to the top of the Empire State Building, our hypothetical Holmes listened and smelled for planes, as strange as it seemed? Attention is about every one of your senses: sight, smell, hearing, taste, touch. It is about taking in as much as we possibly can, through all of the avenues available to us. It is about learning not to leave *anything* out—anything, that is, that is relevant to the goals that you've set. And it is about realizing that all of our senses affect us—and will affect us whether or not we are aware of the impact.

To observe fully, to be truly attentive, we must be inclusive and not let anything slide by—and we must learn how our attention may shift without our awareness, guided by a sense that we'd thought invisible. That jasmine? Holmes smelled the letter deliberately. In so doing, he was able to observe the presence of a female influence, and a particular female at that. If Watson had picked up the letter, we can be sure he would have done no such thing. But his nose may very well have grasped the scent even without his awareness. What then?

When we smell, we remember. In fact, research has shown that the memories associated with smell are the most powerful, vivid, and emotional of all our recollections. And what we smell affects what we remember, how we subsequently feel, and what we might be inclined to think as a result. But smell is often referred to as the invisible sense: we regularly experience it without consciously registering it. A smell enters our nose, travels to our olfactory bulb, and makes its way directly to our hippocampus, our amygdala (an emotion-processing center), and our olfactory cortex (which not only deals with smells but is involved in complex memory, learning, and decision-making tasks), triggering a host of thoughts, feelings, and recollections—yet more likely than not, we note neither smell nor memory.

What if Watson, in all of his multiple-continent-spanning womanizing, happened to have dated a woman who wore a jasmine perfume? Let's imagine the relationship a happy one. All of a sudden he may have found himself seeing with added clarity (remember, happy moods equal wider sight), but he may also have failed to note select details because of a certain rosy glow to the whole thing. Maybe the letter isn't so sinister. Maybe Henry isn't in all that much danger. Maybe it would be better to go have a drink and meet some lovely ladies—after all, ladies are lovely, aren't they? And off we go.

And if the relationship had been violent, brutish, and short? Tunnel vision would have set in (bad mood, limited sight), and along with it a brushing aside of most of the elements of note. *Why should that matter? Why should I work harder? I am tired; my senses are overloaded; and I deserve a break. And why is Henry bothering us anyway with this nonsense? Paranormal dog, my foot. I've about had it.*

When we are being inclusive, we never forget that all of our senses are constantly in play. We don't let them drive our emotions and decisions. Instead, we actively enlist their help—as Holmes does with both boot and letter—and learn to control them instead.

In either of the Watson scenarios above, all of the doctor's actions, from the moment of smelling the jasmine, will have been affected. And while the precise direction of the effect is unknowable, one thing is certain. Not only would he have failed to be inclusive in his attention, but his attention will have been hijacked by the eponymous System Watson into a subjectivity that will be all the more limited for its unconscious nature.

It may seem like I'm exaggerating, but I assure you, sensory influences—especially olfactory ones—are a powerful lot. And if we aren't aware of them altogether, as so often happens, they can threaten to take over the carefully cultivated goals and objectivity that we've been working on.

Smell may be the most glaring culprit, but it is far from alone. When we see a person, we are likely to experience the activation of any number of stereotypes associated with that person—though we won't realize it. When we touch something warm or cold, we may become likewise warm or cold in our disposition; and if we are touched by someone in a reassuring way, we may suddenly find ourselves taking more risk or being more confident than we otherwise would. When we hold something heavy, we are more likely to judge something (or someone) to be weightier and more serious. None of this has anything to do with observation and attention per se, except that it can throw us off a carefully cultivated path without our awareness. And that is a dangerous thing indeed.

We don't have to be a Holmes and learn to tell apart hundreds of smells from a single whiff in order to let our senses work for us, to allow our awareness to give us a fuller picture of a scene that we would otherwise have. A scented note? You don't need to know the smell to realize that it is there—and that it might be a potential clue. If you hadn't paid attention to the fragrance, you would have missed the clue's presence altogether—but you may have had your objectivity undermined nevertheless without even being aware of what has taken place. A missing boot? Another missing boot? Maybe it's about a quality other than the boot's appearance—after all, it's the old and ugly one that eventually disappeared for good. You don't need to know much to realize that there may be another sensory clue here that would again be missed if you had forgotten about your other senses. In both cases, a failure to use all senses equals a scene not seen to its full potential, attention that has not been allocated properly, and subconscious cues that color the attention that *is* allocated in a way that may not be optimal.

If we actively engage each of our senses, we acknowledge that the world is multidimensional. Things are happening through our eyes, our nose, our ears, our skin. Each of those senses should rightly tell us something. And if it doesn't, that should also tell us something: that a sense is missing. That something *lacks* smell, or is silent, or is otherwise absent. In other words, the conscious use of each sense can go beyond illuminating the present part of scene and show instead that part of a situation that is often forgotten: that which *isn't* there, which is not present in the environment where by every rightful metric it should be. And absence can be just as important and just as telling as presence.

Consider the case of Silver Blaze, that famous missing racehorse that no one can track down. When Holmes has had a chance to examine the premises, Inspector Gregson, who has failed to find something as seemingly impossible to miss as a horse, asks, "Is there any point to which you would wish to draw my attention?" Why yes, Holmes responds, "To the curious incident of the dog in the night-time." But, protests the inspector, "The dog did nothing in the night-time." To which Holmes delivers the punch line: "That was the curious incident."

For Holmes, the absence of barking is the turning point of the case: the dog must have known the intruder. Otherwise he would have made a fuss.

For us, the absence of barking is something that is all too easy to forget. All too often, we don't even dismiss things that aren't there; we don't remark on them to begin with—especially if the thing happens to be a sound, again a sense that is not as natural a part of attention and observation as sight. But often these missing elements are just as telling and just

as important—and would make just as much difference to our thinking—
as their present counterparts.

We need not be dealing with a detective case for absent information
to play an important role in our thought process. Take, for example, a
decision to buy a cell phone. I'm going to show you two options, and I
would like you to tell me which of them you would rather purchase.

Phone A	Phone B
Wi-fi: 802.11 b/g	802.11 b/g
Talk time: 12 hrs	16 hrs
Standby time: 12.5 days	14.5 days
Memory: 16.0 GB	32.0 GB
Cost: $100	$150

Did you make a decision? Before you read on, jot down either Phone
A or Phone B. Now I'm going to describe the phones one more time. No
information has been changed, but some has been added.

Phone A	Phone B
Wi-fi: 802.11 b/g	802.11 b/g
Talk time: 12 hrs	16 hrs
Standby time: 12.5 days	14.5 days
Memory: 16.0 GB	32.0 GB
Cost: $100	$150
Weight: 135g	300g

Which phone would you rather purchase now? Again, write down your
answer. I'm now going to present the options a third time, again adding
one new element.

Phone A	Phone B
Wi-fi: 802.11 b/g	802.11 b/g
Talk time: 12 hrs	16 hrs
Standby time: 12.5 days	14.5 days
Memory: 16.0 GB	32.0 GB
Cost: $100	$150
Weight: 135g	300g
Radiation (SAR): 0.79 W/kg	1.4 W/kg

Now, which of the two would you prefer?

Chances are, somewhere between the second and third lists of data, you switched your allegiance from Phone B to Phone A. And yet the two phones didn't change in the least. All that did was the information that you were aware of. This is known as omission neglect. We fail to note what we do not perceive up front, and we fail to inquire further or to take the missing pieces into account as we make our decision. Some information is always available, but some is always silent—and it will remain silent unless we actively stir it up. And here I used only visual information. As we move from two to three dimensions, from a list to the real world, each sense comes into play and becomes fair game. The potential for neglecting the omitted increases correspondingly—but so does the potential for gleaning more about a situation, if we engage actively and strive for inclusion.

Now let's go back to that curious dog. He could have barked or not. He didn't. One way to look at that is to say, as the inspector does, he did nothing at all. But another is to say, as Holmes does, that the dog actively chose not to bark. The result of the two lines of reasoning is identical: a silent dog. But the implications are diametrically opposed: passively doing nothing, or actively doing something.

Nonchoices are choices, too. And they are very telling choices at that. Each nonaction denotes a parallel action; each nonchoice, a parallel choice; each absence, a presence. Take the well-known default effect: more often than not, we stick to default options and don't expend the energy to change, even if another option is in fact better for us. We don't choose to contribute to a retirement fund—even if our company will match the contributions—unless the default is set up for contributing. We don't become organ donors unless we are by default considered donors. And the list goes on. It's simply easier to do nothing. But that doesn't mean we've actually not done anything. We have. We've chosen, in a way, to remain silent.

To pay Attention means to pay attention to it all, to engage actively, to use all of our senses, to take in everything around us, including those things that don't appear when they rightly should. It means asking questions and making sure we get answers. (Before I even go to buy that car or cell phone, I should ask: what are the features I care about most? And then I should be sure that I am paying attention to those features—and not to something else entirely.) It means realizing that the world is three-dimensional and multi-sensory and that, like it or not, we will be influenced by our environment, so our best bet is to take control of that influence by paying attention to everything that surrounds us. We may not be able to emerge with the entire situation in hand, and we may end up making a choice that, upon further reflection, is not the right one after all. But it won't be for lack of trying. All we can do is observe to the best of our abilities and never assume anything, including that absence is the same as nothing.

4. Be Engaged

Even Sherlock Holmes makes the occasional mistake. But normally these are mistakes of misestimation—of a person, in the case of Irene Adler; a horse's ability to stay hidden in "Silver Blaze"; a man's ability to stay the same in "The Case of the Crooked Lip." It is rare indeed that the mistake is a more fundamental one: a failure of engagement. Indeed, it is only on one occasion, as far as I'm aware, that the great detective is negligent in embodying that final element of attentiveness, an active, present interest and involvement, an engagement in what he is doing—and it almost costs him his suspect's life.

The incident takes place toward the end of "The Stock Broker's Clerk." In the story, the clerk of the title, Hall Pycroft, is offered a position as the business manager of the Franco-Midland Hardware Company by a certain Mr. Arthur Pinner. Pycroft has never heard of the firm and is slated to begin work the following week at a respected stockbrokerage—but the pay is simply too good to pass up. And so he agrees to begin work the next day. His suspicions are aroused, however, when his new employer, Mr. Pinner's brother Harry, looks suspiciously like Mr. Arthur. What's more, he finds that his so-called office employs no other man and doesn't even have a sign on the wall to alert potential visitors of its existence. To top it off, Pycroft's task is nothing like that of a clerk: he is to copy listings out of a thick phone book. When, a week later, he sees that Mr. Harry has the same gold tooth as did Mr. Arthur, he can stand the strangeness no more and so sets the problem before Sherlock Holmes.

Holmes and Watson proceed to accompany Hall Pycroft to the Midlands, to the office of his employer. Holmes thinks he knows just what has gone on, and the plan is to visit the man on the pretense of looking for work, and then confront him as Holmes is wont to do. Every detail is in place. Every aspect of the situation is clear to the detective. It's not like those cases where he actually needs the criminal to fill in major blanks. He knows what to expect. The only thing he requires is the man himself.

But when the trio enters the offices, Mr. Pinner's demeanor is not at all as expected. Watson describes the scene.

> At the single table sat the man whom we had seen in the street, with his evening paper spread out in front of him, and as he looked up at us it seemed to me that I had never looked upon a face which bore such marks of grief, and of something beyond grief—of a horror such as comes to few men in a lifetime. His brow glistened with perspiration, his cheeks were of the dull, dead white of a fish's belly, and his eyes were wild and staring. He looked at his clerk as though he failed to recognize him, and I could see by the astonishment depicted upon our conductor's face that this was by no means the usual appearance of his employer.

But what happens next is even more unexpected—and threatens to foil Holmes's plans entirely. Mr. Pinner attempts to commit suicide.

Holmes is at a loss. This he had not anticipated. Everything up to then is "clear enough, but what is not so clear is why at the sight of us the rogue should instantly walk out of the room and hang himself," he says.

The answer comes soon enough. The man is revived by the good Dr. Watson and provides it himself: the paper. He had been reading a newspaper—or rather, something quite specific in that paper, something that has caused him to lose his emotional equilibrium entirely—when he was interrupted by Sherlock and company. Holmes reacts to the news with uncharacteristic vigor. "'The paper! Of course!' yelled Holmes in a paroxysm of excitement. 'Idiot that I was! I thought so much of our visit that the paper never entered my head for an instant.'"

The moment the paper is mentioned, Holmes knows at once what it means and why it had the effect that it did. But why did he fail to note it in the first place, committing an error that even Watson would have hung his head in shame at making? How did the System Holmes machine become . . . a System Watson? Simple. Holmes says it himself: he had lost interest in the case. In his mind, it was already solved, down to the last detail—the visit, of which he thought so much that he decided it would be fine to disengage from everything else. And that's a mistake he doesn't normally make.

Holmes knows better than anyone else how important engagement is for proper observation and thought. Your mind needs to be active, to be involved in what it's doing. Otherwise, it will get sloppy—and let pass a crucial detail that almost gets the object of your observation killed. Motivation matters. Stop being motivated, and performance will drop off, no matter how well you've been doing up until the end—even if you've successfully done everything you should have been doing up to now, the moment motivation and involvement flag, you slip up.

When we are engaged in what we are doing, all sorts of things happen. We persist longer at difficult problems—and become more likely to solve them. We experience something that psychologist Tory Higgins refers to as flow, a presence of mind that not only allows us to extract more from whatever it is we are doing but also makes us *feel* better and happier: we derive actual, measurable hedonic value from the strength of our active involvement in and attention to an activity, even if the activity is as boring as sorting through stacks of mail. If we have a reason to do it, a reason that engages us and makes us involved, we will both do it better and feel happier as a result. The principle holds true even if we have to expand significant mental effort—say, in solving difficult puzzles. Despite the exertion, we will still feel happier, more satisfied, and more in the zone, so to speak.

What's more, engagement and flow tend to prompt a virtuous cycle of sorts: we become more motivated and aroused overall, and, consequently, more likely to be productive and create something of value. We even become less likely to commit some of the most fundamental errors of observation (such as mistaking a person's outward appearance for factual detail of his personality) that can threaten to throw off even the best-laid plans of the aspiring Holmesian observer. In other words, engagement stimulates System Holmes. It makes it more likely that System Holmes will step up, look over System Watson's shoulder, place a reassuring hand on it, and say, just as it's about to leap into action, *Hold off a minute. I think we should look at this more closely before we act.*

To see what I mean, let's go back for a moment to Holmes—specifically, to his reaction to Watson's overly superficial (and unengaged) judgment of their client in "The Adventure of the Norwood Builder." In the story, Dr. Watson demonstrates a typical System Watson approach to observation: judging too quickly from initial impressions and failing to correct for the specific circumstances involved. Though in this particular case the judgment happens to be about a person—and as it applies to people, it has a specific name: the correspondence bias, a concept we've already encountered—the process it illustrates goes far beyond person perception.

After Holmes enumerates the difficulties of the case and stresses the importance of moving quickly, Watson remarks, "Surely the man's appearance would go far with any jury?" Not so fast, says Holmes. "That is a dangerous argument, my dear Watson. You remember that terrible murderer, Bert Stevens, who wanted us to get him off in '87? Was there ever a more mild-mannered, Sunday school young man?" Watson has to agree that it is, in fact, so. Many times, people are not what they may initially be judged to be.

Person perception happens to be an easy illustration of the engagement process in action. As we go through the following steps, realize that they apply to anything, not just to people, and that we are using people merely to help us visualize a much more general phenomenon.

The process of person perception is a deceptively straightforward one. First, we categorize. What is the individual doing? How is he acting? How does he *appear*? In Watson's case, this means thinking back to John Hector McFarlane's initial entrance to 221B. He knows at once (by Holmes's prompting) that their visitor is a solicitor and a Freemason—two respectable occupations if ever there were any in nineteenth-century London. He then notes some further details.

> He was flaxen-haired and handsome, in a washed-out negative fashion, with frightened blue eyes, and a clean-shaven face, with a weak, sensitive mouth. His age may have been about twenty-seven, his dress and bearing that of a gentleman. From the pocket of his light summer overcoat protruded the bundle of endorsed papers which proclaimed his profession.

(Now imagine this process happening in the exact same way for an object or location or whatever else. Take something as basic as an apple. Describe it: how does it look? Where is it? Is it doing anything? Even sitting in a bowl is an action.)

After we categorize, we characterize. Now that we know what he's doing or how he seems, what does that imply? Are there some underlying traits or characteristics that are likely to have given rise to my initial impression or observation? This is precisely what Watson does when he tells Holmes, "Surely the man's appearance would go far with any jury." He has taken the earlier observations, loaded as they might be—handsome, sensitive, gentlemanly bearing, papers proclaiming his profession as a solicitor—and decided that taken together, they imply trustworthiness. A solid, straightforward nature that no jury could doubt. (Think you can't characterize an apple? How about inferring healthiness as an intrinsic characteristic because the apple happens to be a fruit, and one that appears to have great nutritional value given your earlier observations?)

Finally, we correct: *Is there something that may have caused the action other than my initial assessment (in the characterization phase)? Do I need to adjust my initial impressions in either direction, augmenting some elements or discounting others?* That sounds easy enough: take Watson's judgment of trustworthiness, or your judgment of healthiness, and see if it needs to be adjusted.

Except, there's one major problem: while the first two parts of the process are nearly automatic, the last is far less so—and often never happens at all. Consider that in the case of John McFarlane, it is not Watson who corrects his impression. He takes it for what it is and is about to move on. Instead, it is the ever-engaged Holmes who points out that Watson's reasoning "is a dangerous argument." McFarlane may or may not be able to rely on his appearance to go far with any jury. It all depends on the jury—and on the other arguments of the case. Appearance alone can be deceptive. What can you really tell about McFarlane's trustworthiness from simply looking at him? Back to that apple: can you really know it is healthy by examining its exterior? What if this particular apple is not only *not* organic, but has come from an orchard that is known to use illegal pesticides—and has not been properly washed or handled since? Appearances can deceive even here. Because you already have a schema of an apple set in your mind, you may deem it too time consuming and unnecessary to go any further.

Why do we so often fail at this final stage of perception? The answer lies in that very element we were discussing: engagement.

Perception comes in two flavors, passive and active, and the distinction is not the one you might think. In this case, System Watson is the active one, System Holmes, the passive. As passive perceivers, we just observe. And by that I mean that we do not do anything else. We are not, in other words, multitasking. Holmes the passive observer focuses all of his faculties on the subject of observation, in this case, John Hector McFarlane. He listens, as is his habit, "with closed eyes and fingertips together." The word *passive* can be misleading, in that there is nothing passive about his concentrated perception. What is passive is his attitude to the rest of the world. He will not be distracted by any other task. As passive observers, we are not doing anything else; we are focused on observing. A better term in my mind would be engaged passivity: a state that is the epitome of engagement but happens to be focused on only one thing, or person, as the case may be.

In most situations, however, we don't have the benefit of simply observing (and even when we do, we don't often choose to do so). When we are in a social environment, which defines most situations, we can't just stand back and watch. Instead, we are in a state of de facto multitasking, trying to navigate the complexities of social interaction at the same time as we make attributional judgments, be it about people, things, or environments. Active perception doesn't mean active in the sense of present and engaged. Active perception means that the perceiver is, literally, active: doing many things at once. Active perception is System Watson trying to run all over the place and not miss a thing. It is the Watson who not only examines his visitor, but worries about the doorbell, the newspaper, when lunch will be served, how Holmes is feeling, all in the same moment. A better term here would be disengaged activity: a state where you seem to be active and productive, but are actually doing nothing to its fullest potential, spreading thin your attentional resources.

What separates Holmes from Watson, the passive observer from the active one, engaged passivity from disengaged activity, is precisely the descriptor I've used in both cases: engagement. Flow. Motivation. Interest. Call it what you may. That thing that keeps Holmes focused exclusively on his visitor, that enraptures him and prevents his mind from wandering anywhere but to the object at hand.

In a set of classic studies, a group of Harvard researchers set out to demonstrate that active perceivers categorize and characterize on a near-subconscious level, automatically and without much thought, but then fail to implement the final step of correction—even when they have all of the information to do so—and so end up with an impression of someone that does not take into account all of the variables of the interaction. Like Watson, they remember only that a jury would like a man's appearance; unlike Holmes, they fail to take into account those factors that might make that appearance a deceptive one—or those circumstances under which a jury would dismiss any appearance, no matter how trustworthy, as false (like additional evidence so weighty it renders all subjective aspects of the case largely irrelevant).

In the first study, the researchers tested whether individuals who were cognitively "busy," or multitasking in the way that we often are when we juggle numerous elements of a situation, would be able to correct initial impressions by making the necessary adjustment. A group of participants was asked to watch a series of seven video clips in which a woman was having a conversation with a stranger. The clips did not have sound, ostensibly to protect the privacy of those speaking, but did include subtitles at the bottom of each clip that told participants the topic of conversation. In five of the seven videos, the woman behaved in an anxious fashion, while in the other two she remained calm.

While everyone watched the exact same videos, two elements differed: the subtitles and the task that the participants were expected to perform. In one condition, the five anxious clips were paired with anxiety-provoking topics, such as sex life, while in the other, all seven clips were paired with neutral topics like world travel (in other words, the five clips of anxious behavior would seem incongruous given the relaxing subject). And within each of these conditions, half of the participants were told that they would be rating the woman in the video on some personality dimensions, while the other half was expected to both rate personality and be able to recall the seven topics of conversation in order.

What the researchers found came as no shock to them, but it did shake up the way person perception—the way we view other people—had always been seen. While those individuals who had to focus only on the woman adjusted for the situation, rating her as dispositionally more anxious in the neutral topic condition and as less anxious in the anxiety-inducing topic condition, those who had to recall the conversation topics completely failed to take those topics into account in their judgment of the woman's anxiety. They had all of the information they needed to make the judgment, but they never thought to use it. So even though they knew that the situation would make anyone anxious *in theory*, in practice they simply decided that the woman was a generally anxious person. What's more, they predicted that she would continue to be anxious in future scenarios, regardless of how anxiety-provoking those scenarios were. And the better they recalled the topics of conversation, the more extremely off their predictions were. In other words, the busier their brains were, the less they adjusted after forming an initial impression.

The news here is both good and bad. First, the obviously bad: in most situations, under most circumstances, we are active observers, and as such, more likely than not to make the error of unconsciously, automatically categorizing and characterizing, and then failing to correct that initial impression. And so we go by appearances; we forget to be subtle; we forget how easily a person can be influenced at any given point by myriad forces, internal and external. Incidentally, this works whether or not you tend, as most Westerners do, to infer stable traits over passing states, or, as many Eastern cultures do, to infer states over traits; whatever direction you err in, you will fail to adjust.

But there's good news. Study after study shows that individuals who are motivated correct more naturally—and more correctly, so to speak—than those who are not. In other words, we have to both realize that we tend to form autopilot-like judgments and then fail to adjust them, and we have to *want*, actively, to be more accurate. In one study, psychologist Douglas Krull used the same initial setup as the Harvard anxiety research—but gave some participants an additional goal: estimate the amount of anxiety caused by the interview questions. Those who regarded the situation were far less likely to decide that the woman was simply an anxious person—even when they were busy with the cognitive rehearsal task.

Or, let's take another commonly used paradigm: the political statement that is assigned to a subject rather than deliberately chosen. Take capital punishment (since we've mentioned that same issue in the past, and it fits nicely into Holmes's criminal world; it's also often used in these experimental settings). Now, you might have one of three, broadly speaking, attitudes toward the death penalty: you might be for it, you might be against it, or you might not particularly care, or not really know, or have never really given it much thought. If I were to give you a brief article with arguments that support capital punishment, how would you respond to it?

The answer is, it depends. If you don't particularly know or care one way or the other—if you are more disinterested or disengaged—you are more likely than not to take the article at something like face value. If you have no real reason to doubt the source and it seems logical enough, you are likely to let it persuade you. You will categorize and characterize, but there will be little need for correction. Correction takes effort, and you have no personal reason to exert any. Contrast this with your reaction if you are a passionate opponent/proponent of the death penalty. In either case, you will pay attention at the mere mention of the theme of the article. You will read it much more carefully, and you will expend the effort necessary for correction. The correction may not be the same if you agree as if you disagree—in fact, you may even overcorrect if you oppose the article's points, going too far in the opposite direction—but whatever the case, you will engage much more actively, and you will exert the mental effort that is necessary to challenge your initial impressions. Because it matters to you to get it right.

(I chose a political issue on purpose, to illustrate that the context need not be related to people, but just think what a difference in perception there would be if you met for the first time a random person versus someone you knew was going to be interviewing or somehow evaluating you shortly. In which case are you more likely to be careful about your impressions, lest you be wrong? In which will you expend more effort to correct and recalibrate?)

When you feel strong personal engagement with something, you will feel it is worth that extra push. And if you are engaged in the process itself—in the idea of observing more carefully, being more attentive and alert—you will be that much more likely to challenge yourself to accuracy. Of course, you need to be aware of the process to begin with—but now you are. And if you realize that you should engage but don't feel up to it? Psychologist Arie Kruglanski has spent his career studying a phenomenon known as the Need for Closure: a desire of the mind to come to some definitive knowledge of an issue. Beyond exploring how individuals differ in that need, Kruglanski has demonstrated that we can manipulate it in order to be more attentive and engaged—and to make sure we complete the correction stage in our judgments.

This can be accomplished in several ways. Most effectively, if we are made to feel accountable in our judgments, we will spend more time looking at angles and possibilities before making up our minds—and so will expend the correctional effort on any initial impressions, to make sure they are accurate. Our minds won't "close" (or, as Kruglanski calls it, "freeze") in their search until we are fairly sure we've done all we can. While there isn't always an experimenter there to hold us accountable, we can do it for ourselves by setting up each important judgment or observation as a challenge. *How accurate can I be? How well can I do? Can I improve my ability to pay attention over the last time?* Such challenges not only engage us in the task of observation and make it more intrinsically interesting, but they also make us less likely to jump to conclusions and issue judgments without a lot of prior thought.

The active observer is hampered because he is trying to do too many things at once. If he is in a social psychology experiment and forced to remember seven topics in order, or a string of digits, or any number of things that psychologists like to use to ensure cognitive busyness, he is basically doomed. Why? Because the experiments are forcibly preventing engagement. You cannot engage—unless you have eidetic memory or have read up on your memory palace skills—if you are trying desperately to remember unrelated information (actually, even if it's related information; the point is, your resources are engaged elsewhere).

But I have news for you: our life is not a social psychology experiment. We are never *required* to be active observers. No one is asking us to recall, in exact order, a conversation or to make a speech of which we hadn't been aware previously. No one is forcing us to limit our engagement. The only ones that do that is us, ourselves. Be it because we've lost interest, as Holmes did with Mr. Pycroft's case, or because we're too busy thinking about a jury trial in the future to focus on the man in the present, like Watson, when we disengage from a person or a situation it is our prerogative. We can just as well not do it.

When we want to engage, believe me, we can. And not only will we then make fewer mistakes of perception, but we will become the types of focused, observant people that we may have thought we were incapable of becoming. Even children who have been diagnosed with ADHD can find themselves able to focus on certain things that grab them, that activate and engage their minds. Like video games. Time after time, video games have proven able to bring out the attentional resources in people that they never suspected they had. And what's more, the kind of sustained attention and newfound appreciation of detail that emerges from the process of engagement can then transfer to other domains, beyond the screen. Cognitive neuroscientists Daphné Bavelier and C. Shawn Green, for instance, have found repeatedly that so-called "action" video games— games characterized by high speed, high perceptual and motor load, upredictability, and the need for peripheral processing—enhance visual attention, low-level vision, processing speed, attentional, cognitive, and social control, and a number of other faculties across domains as varied as the piloting of unmanned drones and laparoscopic surgery. The brain can actually change and learn to sustain attention in a more prolonged fashion—and all because of moments of engagement in something that *actually* mattered.

We began the chapter with mind wandering, and that is where we will end it. Mind wandering is anathema to engagement. Be it mind wandering from lack of stimulation, mind wandering from multitasking (basically, most of modern existence), or mind wandering because of a forced laboratory paradigm, it cannot coexist with engagement. And so, it cannot coexist with mindful attention, the Attention that we need for Observation.

And yet we constantly make the active choice to disengage. We listen to our headphones as we walk, run, take the subway. We check our phones when we are having dinner with our friends and family. We think of the next meeting while we are in the current one. In short, we occupy our minds with self-made memorization topics or distracting strings of numbers. The Daniel Gilberts of the world don't need to do it for us. In fact, Dan Gilbert himself tracked a group of over 2,200 adults in their regular days through iPhone alerts, asking them to report on how they were feeling, what they were doing, and whether or not they were thinking of something other than the activity they had been involved in when they received the alert. And you know what he found? Not only do people think about something other than what they're doing about as often as they think about what they are doing—46.9 percent of the time, to be exact—but what they are actually doing doesn't seem to make a difference; minds wander about equally no matter how seemingly interesting and engaging or boring and dull the activity.

An observant mind, an attentive mind, is a present mind. It is a mind that isn't wandering. It is a mind that is actively engaged in whatever it is that it happens to be doing. And it is a mind that allows System Holmes to step up, instead of letting System Watson run around like crazy, trying to do it all and see it all.

I know a psychology professor who turns off her email and Internet access for two hours every day, to focus exclusively on her writing. I think there's much to learn from that self-enforced discipline and distance. It's certainly an approach I wish I took more often than I do. Consider the results of a recent nature intervention by a neuroscientist who wanted to demonstrate what could happen if people took three days to be completely wireless in the wild: creativity, clarity in thought, a reboot of sorts of the brain. We can't all afford a three-day wilderness excursion, but maybe, just maybe, we can afford a few hours here and there where we can make a conscious choice: focus.

SHERLOCK HOLMES FURTHER READING

"I noticed that [his hand] was all mottled over . . ." "You have been in Afghanistan, I perceive." from *A Study in Scarlet*, chapter 1: Mr. Sherlock Holmes, p. 7.

"I knew you came from Afghanistan." "Before turning to those moral and mental aspects . . ." from *A Study in Scarlet*, chapter 2: The Science of Deduction, p. 15.

"What does Dr. James Mortimer, the man of science, ask of Sherlock Holmes, the specialist in crime?" from *The Hound of the Baskervilles*, chapter 1: Mr. Sherlock Holmes, p. 5.

"My body has remained in this armchair . . ." from *The Hound of the Baskervilles*, chapter 3: The Problem, p. 22.

"Let us continue our reconstruction." from *The Return of Sherlock Holmes*, "The Adventure of the Priory School," p. 932.

"*It may possibly recur to your memory that when I examined the paper upon which the printed words were fastened . . .*" from *The Hound of the Baskervilles,* chapter 15: A Retrospection, p. 156.

"*Is there any point to which you would wish to draw my attention?*" from *The Memoirs of Sherlock Holmes,* "Silver Blaze," p. 1.

"*At the single table sat the man whom we had seen in the street . . .*" from *The Memoirs of Sherlock Holmes,* "The Stockbroker's Clerk," p. 51.

"*Surely the man's appearance would go far with any jury?*" from *The Return of Sherlock Holmes,* "The Adventure of the Norwood Builder," p. 829.

CHAPTER FOUR

Exploring the Brain Attic:
The Value of Creativity and Imagination

A young solicitor, John Hector McFarlane, wakes up one morning to find his life upended: overnight he has become the single most likely suspect in the murder of a local builder. He barely has time to reach Sherlock Holmes to tell his story before he is swept off to Scotland Yard, so damning is the evidence against him.

As he explains to Holmes before he is whisked away, he had first met the victim, a certain Jonas Oldacre, only the prior afternoon. The man had arrived at McFarlane's offices and asked him to copy and witness his will—and to Mr. McFarlane's surprise, that will left him all of the builder's property. He was childless and alone, explained Oldacre. And once upon a time, he had known McFarlane's parents well. He wanted to commemorate the friendship with the inheritance—but, he urged, McFarlane was not to breathe a word of the transaction to his family until the following day. It was to be a surprise.

That evening the builder asked the solicitor to join him for dinner, so that they might afterward go over some important documents in connection with the estate. McFarlane obliged. And that, it seems, was that. Until, that is, the following morning's papers described Oldacre's death—and the burning of his body in the timber yard at the back of his house. The most likely suspect: young John Hector McFarlane, who not only stood to inherit the dead man's estate, but had also left his walking stick (bloodied) at the scene of the crime.

McFarlane is summarily arrested by Inspector Lestrade, leaving Holmes with his strange tale. And though the arrest seems to make sense—the inheritance, the stick, the nighttime visit, all the indications that point to McFarlane's guilt—Holmes can't help but feel that something is off. "I know it's all wrong," Holmes tells Watson. "I feel it in my bones."

Holmes's bones, however, are in this instance going against the preponderance of evidence. As far as Scotland Yard is concerned, the case is as close to airtight as they come. All that remains is to put the final touches on the police report. When Holmes insists that all is not yet clear, Inspector Lestrade begs to differ. "Not clear? Well, if that isn't clear, what could be clear?" he interjects.

> "Here is a young man who learns suddenly that, if a certain older man dies, he will succeed to a fortune. What does he do? He says nothing to anyone, but he arranges that he shall go out on some pretext to see his client that night. He waits until the only other person in the house is in bed, and then in the solitude of a man's room he murders him, burns his body in the wood-pile, and departs to a neighbouring hotel."

As if that weren't enough, there's more: "The blood-stains in the room and also on the stick are very slight. It is probable that he imagined his crime to be a bloodless one, and hoped that if the body were consumed it would hide all traces of the method of his death—traces which, for some reason, must have pointed to him. Is not all this obvious?"

Holmes remains unconvinced. He tells the inspector:

> "It strikes me, my good Lestrade, as being just a trifle too obvious. You do not add imagination to your other great qualities, but if you could for one moment put yourself in the place of this young man, would you choose the very night after the will had been made to commit your crime? Would it not seem dangerous to you to make so very close a relation between the two incidents? Again, would you choose an occasion when you are known to be in the house, when a servant has let you in? And, finally, would you take the great pains to conceal the body, and yet leave your own stick as a sign that you were the criminal? Confess, Lestrade, that all this is very unlikely."

But Lestrade just shrugs his shoulders. What does imagination have to do with it? Observation and deduction, sure: these are the lynchpins of detective work. But imagination? Isn't that just a flimsy retreat of the less hard-minded and scientific professions, those artistic dalliers who couldn't be further from Scotland Yard?

Lestrade doesn't understand just how wrong he is—and just how central a role imagination plays, not just to the successful inspector or detective but to any person who would hold himself as a successful thinker. If he were to listen to Holmes for more than clues as to a suspect's identity or a case's line of inquiry, he would find that he might have less need of turning to him in the future. For, if imagination does not enter into the picture—and do so *before* any deduction takes place—all of those observations, all of that understanding of the prior chapters will have little value indeed.

Imagination is the essential next step of the thought process. It uses the building blocks of all of the observations that you've collected to create the material that can then serve as a solid base for future deduction, be it as to the events of that fateful Norwood evening when Jonas Oldacre met his death or the solution to a pesky problem that has been gnawing at you at home or at work. If you think that you can skip it, that it is something unscientific and frivolous, you'll find yourself having wasted much effort only to arrive at a conclusion that, as clear and obvious as it may seem to you, could not be further from the truth.

What is imagination, and why is it so important? Why, of all things to mention to Lestrade, does Holmes focus on this particular feature, and what is it doing in something as strict-sounding as the scientific method of the mind?

Lestrade isn't the first to turn his nose up at the thought of imagination playing a role in good old scientific reason, nor is Holmes alone in his insistence to the contrary. One of the greatest scientific thinkers of the twentieth century, Nobel-winning physicist Richard Feynman, frequently voiced his surprise at the lack of appreciation for what he thought was a central quality in both thinking and science. "It is surprising that people do not believe that there is imagination in science," he once told

an audience. Not only is that view patently false, but "it is a very interesting kind of imagination, unlike that of the artist. The great difficulty is in
trying to imagine something that you have never seen, that is consistent
in every detail with what has already been seen, and that is different from
what has been thought of; furthermore, it must be definite and not a
vague proposition. That is indeed difficult."

It's tough to find a better summation and definition of the role of
imagination in the scientific process of thought. Imagination takes the
stuff of observation and experience and recombines them into something new. In so doing, it sets the stage for deduction, the sifting through
of imaginative alternatives to decide: out of all of the possibilities you've
imagined, which is the definite one that best explains all of the facts?

In imagining, you bring into being something hypothetical, something that may or may not exist in actuality but that you have actively
created in your own mind. As such, what you imagine "is different from
what has been thought of." It's not a restatement of the facts, nor is it a
simple line from A to B that can be drawn without much thought. It is
your own synthesis and creation. Think of imagination as a kind of essential mental space in your attic, where you have the freedom to work
with various contents but don't yet have to commit to any storage or or-
ganizational system, where you can shift and combine and recombine
and mess around at will and not be afraid of disturbing the main attic's
order or cleanliness in any way.

That space is essential in the sense of there not being a functional attic
without it: you can't have a storage space that is filled to the brim with
boxes. How would you ever come inside? Where would you pull out the
boxes to find what you need? How would you even see what boxes were
available and where they might be found? You need space. You need light.
You need to be able to access your attic's contents, to walk inside and look
around and see what is what.

And within that space, there is freedom. You can temporarily place
there all of the observations you've gathered. You haven't yet filed them
away or placed them in your attic's permanent storage. Instead, you lay
them all out, where you can see them, and then you play around. What
patterns emerge? Can something from permanent storage be added to

make a different picture, something that makes sense? You stand in that open space and you examine what you've gathered. You pull out different elements, try out different combinations, see what works and what doesn't, what feels right and what doesn't. And you come away with a creation that is unlike the facts or observations that have fed into it. It has its roots in them, true, but it is its own unique thing, which exists only in that hypothetical state of your mind and may or may not be real or even true.

But that creation isn't coming out of the blue. It is grounded in reality. It is drawing upon all those observations you've gathered up to that point, "consistent in every detail with what has already been seen." It is, in other words, growing organically out of those contents that you've gathered into your attic through the process of observation, mixed with those ingredients that have always been there, your knowledge base and your understanding of the world. Feynman phrases it thus: "Imagination in a tight straightjacket." To him, the straightjacket is the laws of physics. To Holmes, it is essentially the same thing: that base of knowledge and observation that you've acquired to the present time. Never is it simply a flight of fancy; you can't think of imagination in this context as identical to the creativity of a fiction writer or an artist. It can't be. First, for the simple reason that it *is* grounded in the factual reality that you've built up, and second, because it "must be definite and not a vague proposition." Your imaginings have to be concrete. They have to be detailed. They don't exist in reality, but their substance must be such that they could theoretically jump from your head straight into the world with little adjustment. Per Feynman, they are in a straightjacket—or, in Holmes's terms, they are confined and determined by your unique brain attic. Your imaginings must use it as their base and they must play by its rules—and those rules include the observations you've so diligently gathered. "The game is," continues Feynman, "to try to figure out what we know, what's possible? It requires an analysis back, a checking to see whether it fits, it's allowed according to what is known."

And in that statement lies the final piece of the definition. Yes, imagination must come from a basis in real, hard knowledge, from the concreteness and specificity of your attic. And yes, it serves a greater purpose:

a setup for deduction, be it of a scientific truth, a solution to a murder, or a decision or problem in your own life that is far removed from both. And in all these instances, it must deal with certain constraints. But it is also free. It is fun. It is, in other words, a game. It is the most playful part of a serious endeavor. Not for nothing does Holmes utter the famed refrain "The game is afoot," in the opening lines of "The Adventure of the Abbey Grange." That simple phrase conveys not only his passion and excitement but his approach to the art of detection and, more generally, of thought: it is a serious thing indeed, but it never loses the element of play. That element is necessary. Without it, no serious endeavor stands a chance.

We tend to think of creativity as an all-or-nothing, you-have-it-or-you-don't characteristic of the mind. But that couldn't be further from the truth. Creativity *can* be taught. It is just like another muscle—attention, self-control—that can be exercised and grow stronger with use, training, focus, and motivation. In fact, studies have shown that creativity is fluid and that training enables people to become more creative: if you think your imagination can grow with practice, you will become better at imaginative pursuits. (There, again, is that persistent need for motivation.) Believing you can be as creative as the best of them and learning creativity's essential components is crucial to improving your overall ability to think, decide, and act in a way that would more befit a Holmes than a Watson (or a Lestrade).

Here we explore that mind space, that stage for synthesis, recombination, and insight. That deceptively lighthearted arena that will allow Holmes to solve the case of the Norwood builder—for solve it he will; and as you'll see, Lestrade's confidence in the obvious will prove both misguided and short-lived.

Learning to Overcome Imaginative Doubt

Picture the following. You are led into a room with a table. On the table are three items: a box of tacks, a book of matches, and a candle. You are told that you have only one assignment: attach the candle to the wall. You can take as much time as you need. How do you proceed?

If you are like over 75 percent of the participants in the now-classic study by the Gestalt psychologist Karl Duncker, you would likely try one of two routes. You might try to tack the candle onto the wall—but you'll quickly find that method to be futile. Or you might try to light the candle and use the dripping wax to attach it to the wall, foregoing the box of tacks entirely (after all, you might think, it could be a distracter!). Again, you'd fail. The wax is not strong enough to hold the candle, and your contraption will collapse. What now?

For the real solution you need some imagination. No one sees it at once. Some people find it after only a minute or two of thought. Others see it after faltering through several unsuccessful attempts. And others fail to solve it without some outside help. Here's the answer. Take the tacks out of the box, tack the box to the wall, and light the candle. Soften the bottom of the candle with a match, so that the wax begins to drip into the box, and place the candle inside the box, on top of the soft pillow of wax. Secure. Run out of the room before the candle burns low enough to set the box on fire. Voilà.

Why don't so many people see that alternative? They forget that between observation and deduction there lies an important mental moment. They take the hot System Watson route—action, action, action—underestimating the crucial need for the exact opposite: a moment of quiet reflection. And so they understandably go at once for the most natural or most obvious solutions. The majority of people in this situation do not see that something obvious—a box of tacks—might actually be something *less* obvious: a box *and* tacks.

This is known as functional fixedness. We tend to see objects the way they are presented, as serving a specific function that is already assigned. The box and tacks go together as a box of tacks. The box holds the tacks; it does not have another function. To go past that and actually break the object into two component parts, to realize that the box and matches are two different things, takes an imaginative leap (Duncker, coming from the Gestalt school, was studying precisely this question, of our tendency to see the whole over the parts).

Indeed, in follow-ups to Duncker's original study, one experiment showed that if the objects were presented separately, with the tacks sitting beside the box, the percentage of people who solved the problem rose dramatically. Ditto with a simple linguistic tweak: if participants were primed, prior to encountering the candle problem, with a series of words connected with *and* instead of *of,* as in, "a box *and* tacks," they were much more likely to see the solution. And even if the words were just under-lined separately, as five items (candle, book of matches, and box of tacks), participants were also much more likely to solve the problem.

But the original problem requires some thought, a shift away from the obvious without any external help. It's not as simple as looking at every-thing you've observed and right away acting or trying to deduce the most likely scenario that would satisfy your objective. Those people who were able to solve it knew the importance of *not* acting, the value of letting their minds take the situation in and give it some internal, quiet thought. In short, they realized that between observation and deduction lies the crucial, irreplaceable step of imagination.

It's easy to see Sherlock Holmes as a hard, cold reasoning machine: the epitome of calculating logic. But that view of Holmes the Logical Au-tomaton couldn't be further from the truth. Quite the contrary. What makes Holmes who he is, what places him above detectives, inspectors, and civilians alike, is his willingness to engage in the nonlinear, embrace the hypothetical, entertain the conjecture; it's his capacity for creative thought and imaginative reflection.

Why then do we tend to miss this softer, almost artistic side and focus instead on the detective's computer-like powers of rational calculation? Simply put, that view is both easier and safer. It is a line of thinking that is well ingrained into our psychology. We have been trained to do it from an early age. As Albert Einstein put it, "Certainly we should take care not to make the intellect our god; it has, of course, powerful muscles, but no per-sonality. It cannot lead, it can only serve; and it is not fastidious in its choice of a leader." We live in a society that glorifies the computer model, that idolizes the inhuman Holmes, who can take in countless data points as a matter of course, analyze them with startling precision, and spit out a solu-tion. A society that gives short shrift to the power of something as unquan-tifiable as imagination and focuses instead on the power of the intellect.

But wait, you might think, that's completely bogus. We also *thrive* on the idea of innovation and creativity. We are living in the age of the entrepreneur, of the man of ideas, of Steve Jobs and the "Think Different" motto. Well, yes and no. That is, we value creativity on the surface, but in our heart of hearts, imagination can scare us like crazy.

As a general rule, we dislike uncertainty. It makes us uneasy. A certain world is a much friendlier place. And so we work hard to reduce whatever uncertainty we can, often by making habitual, practical choices, which protect the status quo. You know the saying, "Better the devil you know"? That about sums it up.

Creativity, on the other hand, requires novelty. Imagination is all about new possibilities, eventualities that don't exist, counterfactuals, a recombination of elements in new ways. It is about the untested. And the untested is uncertain. It is frightening—even if we aren't aware of just how much it frightens us personally. It is also potentially embarrassing (after all, there's never a guarantee of success). Why do you think Conan Doyle's inspectors are always so loath to depart from standard protocol, to do anything that might in the least endanger their investigation or delay it by even an instant? Holmes's imagination frightens them.

Consider a common paradox: organizations, institutions, and individual decision makers often reject creative ideas even as they state openly that creativity is an important and sometimes central goal. Why? New research suggests that we may hold an unconscious bias against creative ideas much like we do in cases of racism or phobias.

Remember the Implicit Association Test from chapter two? In a series of studies, Jennifer Mueller and colleagues decided to modify it for something that had never appeared in need of testing: creativity. Participants had to complete the same good/bad category pairing as in the standard IAT, only this time with two words that expressed an attitude that was either practical (*functional, constructive,* or *useful*) or creative (*novel, inventive,* or *original*). The result indicated that even those people who had explicitly ranked creativity as high on their list of positive attributes showed an implicit bias against it relative to practicality under conditions of uncertainty. And what's more, they also rated an idea that had been pretested as creative (for example, a running shoe that uses nanotechnology to adjust fabric thickness to cool the foot and reduce blisters) as *less* creative than their more certain counterparts. So not only were they implicitly biased, but they exhibited a failure to see creativity for what it was when directly faced with it.

True, that effect was seen only in uncertain conditions—but doesn't that describe most decision-making environments? It certainly applies to detective work. And corporations. And science. And business. And basically anything else you can think of.

Great thinkers have gotten over that hump, that fear of the void. Einstein had failures. So did Abraham Lincoln, probably one of the few men to go to war a captain and return a private—and to file twice for bankruptcy before assuming the presidency. So did Walt Disney, getting fired from a newspaper for "lack of imagination" (the creativity paradox, if ever there was one, in full force). So did Thomas Edison, inventing over one thousand failed specimens before he came up with a lightbulb that worked. And so did Sherlock Holmes (Irene Adler, anyone? Man with the twisted lip? Or how about that Yellow Face, to which we'll soon return in greater detail?).

What distinguishes them isn't a lack of failure but a lack of fear of failure, an openness that is the hallmark of the creative mind. They may have had that same anticreative bias as most of us at one point in their lives, but one way or another, they managed to squelch it into submission. Sherlock Holmes has one element that a computer lacks, and it is that very element that both makes him what he is and undercuts the image of the detective as nothing more than logician par excellence: imagination.

Who hasn't dismissed a problem because no obvious answer presented itself at once? And which of us hasn't made a wrong decision or taken a wrong turn because we never stopped to think that clear and obvious might be a trifle *too* obvious? Who hasn't persisted in a less-than-ideal setup just because that's the way things were always done—and though better ways may exist, they would depart too much from the tried and true? Better the devil you know.

Our fear of uncertainty keeps us in check when we'd do better to accompany Holmes on one of his imaginative wanderings and play out scenarios that may exist—for the time being, at least—only in our heads.

Einstein, for one, had nothing but intuition to go on when he proposed his grand theory of general relativity. When George Sylvester Viereck asked him, in 1929, whether his discoveries were the result of intuition or inspiration, Einstein replied, "I'm enough of an artist to draw freely on my imagination, which I think is more important than knowledge. Knowledge is limited. Imagination encircles the world." Absent imagination, the great scientist would have been stuck in the certainty of the linear and the easily accessible.

What's more, many problems don't even have an obvious answer to turn to. In the case of our Norwood mystery, Lestrade had a ready-made story and suspect. But what if that didn't exist? What if there was no linear narrative, and the only way to get to the answer was by circuitous and hypothetical meanderings of the mind? (One such case appears in *The Valley of Fear*, when the victim isn't at all who he seems to be—and neither is the house. A lack of imagination in that instance equals a lack of solution.) And in a world far removed from detectives and inspectors and builders, what if there's no obvious job path or better romantic prospect or choice that would make us happier? What if the answer instead requires digging and some creative self-exploration? Not many would change a known devil for an unknown one—and fewer still would exchange it for none at all.

Without imagination we would never be able to reach the heights of thought that we are capable of; we'd be doomed, at the very best, to become very good at spewing back details and facts—but we'd find it difficult to use those facts in any way that could meaningfully improve our judgment and decision making. We'd have an attic stacked with beautifully organized boxes, folders, and materials. And we wouldn't know where to begin to go through them all. Instead, we'd have to thumb through the stacks over and over, maybe finding the right approach, maybe not. And if the right element wasn't there for the taking but had to actually come from two, or even three, different files? Good luck to us.

Let's go back for a moment to the case of the Norwood builder. Why is it that, lacking imagination, Lestrade can't come near solving the mystery and, indeed, comes close to sentencing an innocent man? What does imagination provide here that straightforward analysis does not? Both the inspector and the detective have access to identical information. Holmes doesn't have some secret knowledge that would enable him to see something that Lestrade does not—or at least any knowledge that Lestrade, too, couldn't easily apply in much the same fashion. But not only do the two men choose to use different elements of their shared knowledge; they then interpret what they do know in altogether different lights. Lestrade follows the straightforward approach, and Sherlock a more imaginative one that the inspector does not even conceive to be possible.

At the beginning of the investigation, Holmes and Lestrade start from the exact same point, as John Hector McFarlane gives the entirety of his statement in their joint presence. In fact, it's Lestrade who has an edge of a sort. He has already been to the scene of the crime, while Holmes is only now hearing of it for the first time. And yet, right away, their approaches diverge. When Lestrade, prior to arresting McFarlane and leading him away, asks Holmes whether he has any further questions, Holmes replies, "Not until I have been to Blackheath." Blackheath? But the murder took place in Norwood. "You mean Norwood," Lestrade corrects the detective. "Oh, yes, no doubt that is what I must have meant," replies Holmes, and proceeds, of course, to Blackheath, the home of the unfortunate Mr. McFarlane's parents.

"And why not Norwood?" asks Watson, just as Lestrade had wondered before him.

"Because," replies Holmes, "we have in this case one singular incident coming close to the heels of another singular incident. The police are making the mistake of concentrating their attention upon the second, because it happens to be the one which is actually criminal." Strike one, as you'll see in a moment, against Lestrade's overly straightforward approach.

Holmes is disappointed in his trip. "I tried one or two leads," he tells Watson upon his return, "but could get at nothing which would help our hypothesis, and several points which would make against it. I gave it up at last, and off I went to Norwood." But, as we'll soon see, the time wasn't wasted—nor does Holmes think it was. For, you never know how the most straightforward-seeming events will unfold once you use that attic space of imagination to its fullest potential. And you never know just what piece of information will make a nonsensical puzzle all of a sudden make sense.

Still, the case does not seem to be heading toward a successful resolution. As Holmes tells Watson, "Unless some lucky chance comes our way I fear that the Norwood Disappearance Case will not figure in that chronicle of our successes which I foresee that a patient public will sooner or later have to endure."

And then, from the most unlikely of places, that very lucky chance appears. Lestrade calls it "important fresh evidence" that definitively establishes McFarlane's guilt. Holmes is stricken—until he realizes just what that fresh evidence is: McFarlane's bloody fingerprint on the hallway wall. What to Lestrade is proof positive of guilt to Holmes is the very epitome of McFarlane's innocence. And what's more, it confirms a suspicion that has, to that point, been nothing more than a nagging feeling, an "intuition," as Holmes calls it, that there has been no crime to begin with. Jonas Oldacre is, as a matter of fact, alive and well.

How can that be? How can the exact same piece of information serve, for the inspector, to condemn a man and, for Holmes, to free him—and to cast doubt on the nature of the entire crime? It all comes down to imagination.

Let's go through it step-by-step. First off, there's Holmes's initial response to the story: not to rush immediately to the scene of the supposed crime but rather to acquaint himself with all possible angles, which may or may not prove useful. And so, a trip to Blackheath, to those very parents who are supposed to have known Jonas Oldacre when young and who, of course, know McFarlane. While this may not seem to be particularly imaginative, it does entail a more open-minded and less linear approach than the one espoused by Lestrade: straight to the scene of the crime, and the scene of the crime only. Lestrade has, in a way, closed off all alternate possibilities from the get-go. Why bother to look if everything you need is right in one place?

Much of imagination is about making connections that are not entirely obvious, between elements that may appear disparate at first. When I was younger, my parents gave me a toy of sorts: a wooden pole with a hole in the middle and a ring at the base. Through the hole was threaded a thick string, with two wooden circles on either end. The point of the toy was to get the ring off the pole. It seemed like a piece of cake at first—until I realized that the string with its circles prevented the ring from coming off the obvious way, over the top of the pole. I tried force. And more force. And speed. Maybe I could trick it? I tried to get the string and circles to somehow detach. The ring to slide over the circles that it hadn't slid over in the past. Nothing worked. None of the solutions that seemed most promising were actually solutions at all. Instead, to remove the ring, you had to take a path so circuitous that it took me hours of trying—with days in between—to finally have the patience to reach it. For you had to, in a sense, stop trying to take the ring off. I'd always begun with that ring, thinking that it had to be the right way to go. After all, wasn't the whole point to remove it? It wasn't until I forgot the ring and took a step back to look at the overall picture and to explore its possibilities that I came upon the solution.

I, too, had to go to Blackheath before I could figure out what was going on in Norwood. Unlike Lestrade, I had a strict guide: I would know when I had solved the puzzle correctly. And so I didn't need Holmes's nudging. I realized I was wrong because I would know without a doubt when I was right. But most problems aren't so clear-cut. There's no stubborn ring that gives you only two answers, right and wrong. Instead, there's a whole mass of misleading turns and false resolutions. And absent Holmes's reminder, you may be tempted to keep tugging at that ring to get it off—and think that it has been removed when all you've really done is lodged it farther up the pole.

So, Holmes goes to Blackheath. But that's not the end to his willingness to engage in the imaginative. In order to approach the case of the Norwood builder as the detective does—and accomplish what he accomplishes—you need to begin from a place of open-minded possibility. You cannot equate the most obvious course of events with the only possible course of events. If you do so, you run the risk of never even thinking of any number of possibilities that may end up being the real answer. And, more likely than not, you will fall prey to that nasty confirmation bias that we've seen in play in previous chapters.

In this instance, not only does Holmes hold very real the chance that McFarlane is innocent, but he maintains and plays out a number of hypothetical scenarios that exist only in his mind, whereby each piece of evidence, including the central one of the very death of the builder, is not what it appears to be. In order to realize the true course of events, Holmes must first imagine the possibility of that course of events. Otherwise he'd be like Lestrade, left saying, "I don't know whether you think that McFarlane came out of jail in the dead of the night in order to strengthen the evidence against himself," and following up that seemingly rhetorical statement with, "I am a practical man, Mr. Holmes, and when I have got my evidence I come to my conclusions."

Lestrade's rhetorical certainty is so misplaced *precisely* because he is a practical man who goes straight from evidence to conclusions. He forgets that crucial step in between, that space that gives you time to reflect, to think of other possibilities, to consider what may have occurred, and to follow those hypothetical lines out inside your mind, instead of being forced to use only what is in front of you. (But never underestimate the crucial importance of that observational stage that has come before, the filling up of the staging area with pieces of information for your use: Holmes can come to his conclusions about the thumbprint only because he knows that he did not miss it before. "I know that that mark was not there when I examined the hall yesterday," he tells Watson. He trusts in his observations, in his attention, in the essential soundness of his attic and its contents both. Lestrade, lacking his training and ruled as he is by System Watson, knows no such certainty.)

A lack of imagination can thus lead to faulty action (the arrest or suspicion of the wrong man) *and* to the lack of proper action (looking for the actual culprit). If only the most obvious solution is sought, the correct one may never be found at all.

Reason without imagination is akin to System Watson at the controls. It seems to make sense and it's what we want to do, but it's too impulsive and quick. You cannot possibly assess and see the whole picture—even if the solution ends up being rather prosaic—if you don't take a step back to let imagination have its say.

Consider this counterexample to the conduct of Lestrade. In "The Adventure of Wisteria Lodge," Holmes pays one of his rare compliments to Inspector Baynes: "You will rise high in your profession. You have instinct and intuition." What does Baynes do differently from his Scotland Yard counterparts to earn such praise? He anticipates human nature instead of dismissing it, arresting the wrong man on purpose with the goal of lulling the real criminal into false complacency. (The wrong man, of course, has a preponderance of evidence against him, more than enough for an arrest, and to a Lestrade would seem to be the right man. In fact, Holmes initially mistakes Baynes's arrest as nothing more than a Lestrade-like blunder.) And in this anticipation lies one of the main virtues of an imaginative approach: going beyond simple logic in interpreting facts and instead using that same logic to create hypothetical alternatives. A Lestrade would never think to do something so nonlinear. Why in the world expend the energy to arrest someone if that someone is not who should be arrested according to the law? Lacking imagination, he can think only in a straight line.

In 1968, the high jump was a well-established sport. You would run, you would jump, and you would make your way over a pole in one of several ways. In older days you'd likely use the scissors, scissoring out your legs as you glided over, but by the sixties you'd probably be using the straddle or the belly roll, facing down and basically rolling over the bar. Whichever style you used, one thing was certain: you'd be facing forward when you made your jump. Imagine trying to jump backward. That would be ridiculous.

Dick Fosbury, however, didn't think so. To him, jumping backward seemed like the way to go. All through high school, he'd been developing a backward-facing style, and now, in college, it was taking him higher than it ever had. He wasn't sure why he did it, but if he thought about it, he would say that his inspiration came from the East: from Confucius and Lao Tzu. He didn't care what anyone else was doing. He just jumped with the feeling of the thing. People joked and laughed. Fosbury looked just as ridiculous as they thought he would (and his inspirations sounded a bit ridiculous, too. When asked about his approach, he told *Sports Illustrated*, "I don't even think about the high jump. It's positive thinking. I just let it happen"). Certainly, no one expected him to make the U.S. Olympics team—let alone win the Olympics. But win he did, setting American and Olympic records with his 7-foot-4.25-inch (2.24-meter) jump, only 1.5 inches short of the world record.

With his unprecedented technique, dubbed the Fosbury Flop, Fosbury did what many other more traditional athletes had never managed to accomplish: he revolutionized, in a very real way, an entire sport. Even after his win, expectations were that he would remain a lone bird, jumping in his esoteric style while the rest of the world looked on. But since 1978 no world record has been set by anyone other than a flopper; and by 1980, thirteen of sixteen Olympic finalists were flopping across the bar. To this day, the flop remains the dominant high jump style. The straddle looks old and cumbersome in comparison. Why hadn't anyone thought of replacing it earlier?

Of course, everything seems intuitive in retrospect. But what seems perfectly clear now was completely inventive and unprecedented at the time. No one thought you could possibly jump backward. It seemed absurd. And Fosbury himself? He wasn't even a particularly talented jumper. As his coach, Berny Wagner, put it, "I have a discus thrower who can jump-reach higher than Dick." It was all in the approach. Indeed, Fosbury's height pales in comparison to the current record—8 feet (2.45 meters), held by Javier Sotomayor—and his accomplishment doesn't even break the top twenty. But the sport has never been the same.

Imagination allows us to see things that aren't so, be it a dead man who is actually alive, a way of jumping that, while backward, couldn't be more forward looking, or a box of tacks that can also be a simple box. It lets us see what might have been and what might be even in the absence of firm evidence. When all of the details are in front of you, how do you arrange them? How do you know which are important? Simple logic gets you part of the way there, it's true, but it can't do it alone—and it can't do it without some breathing space.

In our resistance to creativity, we are Lestrades. But here's the good news: our inner Holmes isn't too far away. Our implicit bias may be strong but it's not immutable, and it doesn't need to affect our thinking as much as it does.

Look at the following picture:

Try to connect these dots with three lines, without lifting your pencil from the paper or retracing any of the lines you draw. You must also end the drawing where you began it. You can take up to three minutes.

Have you finished? If you haven't, fear not; you're far from being alone. In fact, you're like 78 percent of study participants who were given the problem to solve. If you have, how long did it take you?

Consider this: if I had turned on a lightbulb in your line of sight while you were working on the problem, you would have been more likely to solve it if you hadn't solved it already—a full 44 percent of people who saw a lit lightbulb solved the puzzle, as contrasted with the 22 percent in the original condition (the one that you just experienced)—and you would have solved it faster than you might have otherwise. The bulb will have activated insight-related concepts in your mind, and in so doing will have primed your mind to think in a more creative fashion than it would as a matter of course. It is an example of priming in action. Because we associate the lightbulb with creativity and insight, we are more likely to persist at difficult problems and to think in a creative, nonlinear fashion when we see it turn on. All of the concepts that are stored in our attic next to the idea of "lightbulb moment" or "insight" or "eureka" become activated, and that activation in turn helps us become more creative in our own approaches.

By the way, here's the solution to the dot problem.

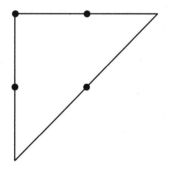

Our natural mindset may well be holding us back, but a simple prime is enough to cue it in a very different direction indeed. And it need not be a lightbulb. Works of art on the walls do the trick, too. The color blue. Pictures of famous creative thinkers. Happy faces. Happy music. (In fact, almost all positive cues.) Plants and flowers and scenes of nature. All of these tend to boost our creativity with or without our awareness. That's cause for celebration.

Whatever the stimulus, as soon as your mind begins to reflect on the idea, you become more likely to embody that very idea. There are even studies that show that wearing a white coat will make you more likely to think in scientific terms and be better at solving problems—the coat likely activates the concept of researchers and doctors, and you begin to take on the characteristics you associate with those people.

But short of lighting bulbs in our blue room with portraits of Einstein and Jobs on the walls while listening to happy music, wearing a white coat, and watering our beautiful roses, how can we best make our way to Holmes's capacity for imaginative thinking?

The Importance of Distance

One of the most important ways to facilitate imaginative thinking, to make sure that we don't move, like Lestrade, straight from evidence to conclusion, is through distance, in multiple senses of the word. In "The Adventure of the Bruce-Partington Plans," a case that comes quite late in the Holmes-Watson partnership, Watson observes:

> One of the most remarkable characteristics of Sherlock Holmes was his power of throwing his brain out of action and switching all his thoughts on to lighter things whenever he had convinced himself that he could no longer work to advantage. I remember that during the whole of that memorable day he lost himself in a monograph which he had undertaken upon the Polyphonic Motets of Lassus. For my own part I had none of this power of detachment, and the day, in consequence appeared to be interminable.

Forcing your mind to take a step back is a tough thing to do. It seems counterintuitive to walk away from a problem that you want to solve. But in reality, the characteristic is not so remarkable either for Holmes or for individuals who are deep thinkers. The fact that it is remarkable for Watson (and that he self-admittedly lacks the skill) goes a long way to explaining why he so often fails when Holmes succeeds.

Psychologist Yaacov Trope argues that psychological distance may be one of the single most important steps you can take to improve thinking and decision making. It can come in many forms: temporal, or distance in time (both future and past); spatial, or distance in space (how physically close or far you are from something); social, or distance between people (how someone else sees it); and hypothetical, or distance from reality (how things might have happened). But whatever the form, all of these distances have something in common: they all require you to transcend the immediate moment in your mind. They all require you to take a step back.

Trope posits that the further we move in distance, the more general and abstract our perspective and our interpretation become; and the further we move from our own perspective, the wider the picture we are able to consider. Conversely, as we move closer once more, our thoughts become more concrete, more specific, more practical—and the closer we remain to our egocentric view, the smaller and more limited the picture that confronts us. Our level of construal influences, in turn, how we evaluate a situation and how we ultimately choose to interact with it. It affects our decisions and our ability to solve problems. It even changes how our brains process information on a neural level (specifically, it tends to engage our prefrontal cortex and medial temporal lobe; more on that later).

In essence, psychological distance accomplishes one major thing: it engages System Holmes. It forces quiet reflection. Distancing has been shown to improve cognitive performance, from actual problem solving to the ability to exercise self-control. Children who use psychological distancing techniques (for example, visualizing marshmallows as puffy clouds, a technique we'll discuss more in the next section) are better able to delay gratification and hold out for a larger later reward. Adults who are told to take a step back and imagine a situation from a more general perspective make better judgments and evaluations, and have better self-assessments and lower emotional reactivity. Individuals who employ distancing in typical problem-solving scenarios emerge ahead of their more immersed counterparts. And those who take a distanced view of political questions tend to emerge with evaluations that are better able to stand the test of time.

You can think of the exercise as a large, complicated puzzle; the box has been lost, so you don't know what exactly you're putting together, and pieces from other similar puzzles have gotten mixed in over the years, so you're not even sure which pieces belong. To solve the puzzle, you must first have a sense of the picture as a whole. Some pieces will jump out right away: the corners, the edges, the colors and patterns that obviously go together. And before you know it, you have a clearer sense of where the puzzle is heading and where and how the remaining pieces should fit. But you'll never solve it if you don't take the time to lay the pieces out properly, identify those telling starter moves, and try to form an image in your mind of the complete picture. Trying to force individual pieces at random will take forever, cause needless frustration, and perhaps lead to your never being able to solve the thing at all.

You need to learn to let the two elements, the concrete, specific pieces (their details and colors, what they tell you, and what they suggest) and the broad, overall picture (the general impression that gives you a sense of the tableau as a whole), work together to help you put the puzzle together. Both are essential. The pieces have been gathered already through close observation; seeing how they fit can be accomplished only by the distance of imagination. It can be any of Trope's distances—temporal, spatial, social, or hypothetical—but distance it must be.

When I was little, I used to love yes-or-no riddle games. One person holds the answer to a simple riddle (one of my favorites as a child: Joe and Mandy are lying on the floor, dead; around them are broken glass, a pool of water, and a baseball. What happened?); the rest try to guess the solution by asking questions that require only a yes or no answer. I could play these for hours and forced many a hapless companion to share the somewhat strange pastime.

Back then I didn't see the riddles as much more than a fun way to pass the time and test my detective prowess—and part of the reason I loved them was because they made me feel up to the task. Only now do I understand fully how ingenious that forced-question method really is: it forces you to separate observation from deduction, whether you want to or not. In a way, the riddles have a built-in road map for how to get to the solution: incrementally, taking frequent breaks to let your imagination consolidate and re-form what it has learned. You can't just barrel on through. You observe, you learn, and you take the time to consider the possibilities, look at the angles, try to place the elements in their proper context, see if you might have come to a mistaken conclusion at an earlier point. The yes-or-no riddle forces imaginative distance. (The solution to Joe and Mandy's dilemma: they are goldfish. The baseball flew in through a window and broke their bowl.)

But absent such an inbuilt cue, how does one go about creating distance? How can one resist Watson's lack of detachment and be able, like Holmes, to know when and how to throw his brain out of action and turn it to lighter things? As it happens, even something as seemingly inborn as creativity and imagination can be broken down into steps that traverse that very you-have-it-or-you-don't divide.

Distancing Through Unrelated Activity

What, pray tell, is a three-pipe problem? It certainly doesn't make it on the list of common problem types in the psychology literature. And yet perhaps it's time it should.

In "The Red-Headed League," Sherlock Holmes is presented with an unusual conundrum, which at first glance has no reasonable solution. Why in the world would someone be singled out for the color of his hair, and then be paid to do nothing but sit around, along with the hair in question, in a closed room for hours on end?

When Mr. Wilson, the man of the flaming-red hair, leaves Holmes after telling his story, Holmes tells Watson that he must give his prompt attention to the matter. "What are you going to do, then?" asks Watson, anxious as ever to know how the case will be resolved. Holmes's reply may come as somewhat of a surprise:

> "To smoke," he answered. "It is quite a three-pipe problem, and I beg that you won't speak to me for fifty minutes." He curled himself

up in his chair, with his thick knees drawn up to his hawk-like nose, and there he sat with his eyes closed and his black clay pipe thrusting out like the bill of some strange bird. I had come to the conclusion that he had dropped asleep, and indeed was nodding myself, when he suddenly sprang out of his chair with the gesture of a man who has made up his mind, and put his pipe down upon the mantel-piece.

A three-pipe problem, then: one that requires doing something other than thinking directly about the problem—i.e., smoking a pipe—in concentrated silence (and, one expects, smoke), for the time that it takes to smoke three pipes. Presumably, one of a subset of problems ranging from the single-pipe problem to the largest number you can smoke without making yourself sick and so putting the entire effort to waste.

Holmes, of course, means something quite a bit more by his response. For him, the pipe is but a means—and one means of many—to an end: creating psychological distance between himself and the problem at hand, so that he can let his observations (in this case, what he has learned from the visitor's story and appearance) percolate in his mind, mixing with all of the matter in his brain attic in leisurely fashion, in order to know what the actual next step in the case should be. Watson would have him do something at once, as suggested by his question. Holmes, however, puts a pipe in between himself and the problem. He gives his imagination time to do its thing undisturbed.

The pipe is but a means to an end, yes, but it is an important, physical means as well. It's significant here that we are dealing with an actual object and an actual activity. A change in activity, to something seemingly unrelated to the problem in question, is one of the elements that is most conducive to creating the requisite distance for imagination to take hold. Indeed, it is a tactic that Holmes employs often and to good effect. He smokes his pipe, but he also plays his violin, visits the opera, and listens to music; these are his preferred distancing mechanisms.

The precise activity isn't as important as its physical nature and its ability to train your thoughts in a different direction. It needs to have several characteristics: it needs to be unrelated to what you are trying to accomplish (if you are solving a crime, you shouldn't switch to solving another crime; if you are deciding on an important purchase, you shouldn't go shopping for something else; and so on); it needs to be something that doesn't take too much effort on your part (if you're trying to learn a new skill, for instance, your brain will be so preoccupied that it won't be able to free up the resources needed to root through your attic; Holmes's violin playing—unless you are, like him, a virtuoso, you need not apply that particular route); and yet it needs to be something that engages you on some level (if Holmes hated pipe smoking, he would hardly benefit from a three-pipe problem; likewise, if he found pipe smoking boring, his mind might be too dulled to do any real thinking, on whatever level—or might find itself unable to detach, in the manner that so afflicts Watson).

When we switch gears, we in effect move the problem that we have been trying to solve from our conscious brain to our unconscious. While we may think we are doing something else—and indeed, our attentional networks become engaged in something else—our brains don't actually stop work on the original problem. We may have left our attic to smoke a pipe or play a sonata, but our staging area remains a place of busy activity, with various items being dragged into the light, various combinations being tried, and various approaches being evaluated.

The key to diagnosing Watson's inability to create distance between himself and a case may well be that he hasn't found a suitably engaging yet not overwhelming activity as a substitute. In some instances he tries reading. Too difficult of a task: not only does he fail to concentrate on the reading, thereby losing the intent of the activity, but he can't stop his mind from returning to the very thing he shouldn't be thinking about. (And yet for Holmes, reading is indeed a suitable distancing method. "Polyphonic Motets of Lassus" anyone?) Other times, Watson tries sitting in contemplation. Too boring, as he himself puts it; he soon finds himself almost nodding off.

In either case, the distancing fails. The mind is simply not doing what it is supposed to—dissociating itself from the present environment and thus engaging its more diffuse attentional network (that same default network that is active when our brains are at rest). It's the opposite of the distraction problem that we encountered in the last chapter. Watson now can't be distracted *enough*. What he should be doing is distracting himself from the case, but instead he is letting the case distract him from his chosen distraction and so failing to get the benefit of either concentrated thought or diffuse attention. Distraction isn't always a bad thing. It all depends on the timing and type. (Interesting fact: we get better at solving insight problems when we are tired or intoxicated. Why? Our executive function is inhibited, so information that would normally be deemed distracting is allowed to filter in. We thus become better at seeing remote associations.) The last chapter was all about mindless distraction; this, on the contrary, is mindful distraction.

But for it to work it's essential to choose the right activity, be it the pipe or the violin or an opera or something else entirely. Something that is engaging enough that it distracts you properly—and yet not so overwhelming that it prevents reflection from taking place in the background. Once you find your sin of choice, you can term the problems and decisions you face accordingly: three-pipe, two-movement, one-museum visit, you get the idea.

In fact, there's one activity that is almost tailor-made to work. And it is a simple one indeed: walking (the very thing that Holmes was doing when he had his insight in "The Lion's Mane"). Walks have been shown repeatedly to stimulate creative thought and problem solving, especially if these walks take place in natural surroundings, like the woods, rather than in more urbanized environments (but both types are better than none—and even walking along a tree-lined street can help). After a walk, people become better at solving problems; they persist longer at difficult tasks; and they become more likely to be able to grasp an insightful solution (like being able to connect those four dots you saw earlier). And all from walking past some trees and some sky.

Indeed, being surrounded by nature tends to increase feelings of well-being, and such feelings, in turn, tend to facilitate problem solving and creative thinking, modulating attention and cognitive control mechanisms in the brain in a way that predisposes us to engage in more Holmes-like imagination. Even the walk can—at times when the pressure seems just too high to handle so that, like Watson, you can't even begin to contemplate doing something else—be forfeited in favor of looking at screen shots of natural scenes. It's not ideal but it just might do the trick in a pinch.

Showers are likewise often associated with imaginative thought, facilitating the same type of distance as Holmes's pipe or a walk in the park. (You can shower for only so long, however. A three-pipe problem would signify quite the shower ahead of you. In such cases, the walk might be the better solution.) Ditto listening to music—Holmes's violin and opera in action—and engaging in visually stimulating activities, such as looking at visual illusions or abstract art.

In every case, that diffuse attentional network is able to do its thing. As our inhibition is lowered, the attentional network takes over whatever is bothering us. It ramps up, so to speak, for whatever comes next. It makes us more likely to grasp remote connections, to activate unrelated memories, thoughts, and experiences that may help in this instance, to synthesize the material that needs to be synthesized. Our unconscious processing is a powerful tool, if only we give it the space and time to work.

Consider a classic problem-solving paradigm known as compound remote associates. Look at these words:

CRAB PINE SAUCE

Now, try to think of a single word that can be added to each of these to form a compound or a two-word phrase.

Done? How long did it take? And how did you come about your solution?

There are two ways to solve this problem. One comes from insight, or seeing the right word after a few seconds of searching, and the other comes from an analytical approach, or trying out word after word until one fits. Here, the proper answer is *apple* (*crab apple, pineapple, applesauce*), and one can arrive at it either by seeing the solution or going through a list of possible candidates (*Cake?* Works for *crab* but not *pine. Grass?* Ditto. Etcetera). The former is the equivalent of picking out those items in the opposite corners of your attic and making them into a third related, yet unrelated, thing that makes complete sense the moment you see it. The

latter is the equivalent of rummaging through your attic slowly and pain-
fully, box by box, and discarding object after object that does not match
until you find the one that does.

Absent imagination, you're stuck with that second not very palatable
alternative, as Watson would be. And while Watson might get to the right
answer eventually in the case of a puzzle like the word associates, in the
real world there's no guarantee of his success, since he doesn't have the
elements laid out in front of him as nicely as those three words, *crab,
pine, sauce.* He hasn't created the requisite mind space for insight to even
be possible. He has no idea which elements may need to come together.
He has, in other words, no conception of the problem.

Even his brain will be different from Holmes's as he approaches the
problem, be it the word association or the case of the builder. At first
glance, if Watson were to come to the right answer on his own, we might
not see an immediate difference. In either Holmes or Watson's case, a
brain scan would show us that a solution has been reached approximately
three hundred milliseconds before the solver realizes it himself. Specifically, we would see a burst of activity from the right anterior temporal
lobe (an area just above his right ear that is implicated in complex cognitive processing), and an increased activation in the right anterior superior temporal gyrus (an area that has been associated with perceiving
emotional prosody—or the rhythm and intonation of language that conveys a certain feeling—and bringing together disparate information in
complex language comprehension).

But Watson may well never reach that point of solution—and we'd
likely know he's doomed long before he himself does. While he's strug-
gling with the puzzle, we would be able to predict if he was heading in the
right direction by looking at neural activity in two areas: the left and
right temporal lobes, associated with the processing of lexical and semantic information, and the mid-frontal cortex, including the anterior
cingulate, associated with attention switching and the detection of in-
consistent and competing activity. That latter activation would be particularly intriguing, as it suggests the *process* by which we're able to gain
insight into a preciously inscrutable problem: the anterior cingulate is
likely waiting to detect disparate signals from the brain, even weak ones

that we are unaware of sending, and turning its attention to them to gain a possible solution, amplifying, so to speak, information that already exists but that needs a little push to be integrated and processed as a general whole. In Watson's brain, we're not likely to see much action. But Holmes's would tell a different story.

In fact, were we to simply compare Watson's brain to Holmes's, we would find telltale signs of Holmes's predisposition to such insights—and Watson's lack thereof—even absent a target for his mind to latch on to. Specifically, we would discover that the detective's brain was more active in the right-hemisphere regions associated with lexical and semantic processing than your average Watson brain, and that it exhibited greater diffuse activation of the visual system.

What would these differences mean? The right hemisphere is more involved in processing such loose or remote associations as often come together in moments of insight, while the left tends to focus on tighter, more explicit connections. More likely than not, the specific patterns that accompany insight signal a mind that is ever ready to process associations that, at first glance, don't seem to be associations at all. In other words, a mind that can find connections between the seemingly unconnected can access its vast network of ideas and impressions and detect even faint links that can then be amplified to recognize a broader significance, if such a significance exists. Insight may seem to come from nowhere, but really, it comes from somewhere quite specific: from the attic and the processing that has been taking place while you've been busy doing other things.

The pipe, the violin, the walk, the concert, the shower, they all have something else in common, beyond the earlier criteria we used to nominate them as good potential activities for creating distance. They allow your mind to relax. They take the pressure off. In essence, all of the mentioned characteristics—unrelated, not too effortful, and yet effortful enough—come together to offer the proper environment for neural relaxation. You can't relax if you're supposed to be working on a problem; hence the unrelatedness. Nor can you relax if you're finding something effortful. And too lax, well, you may not be stimulated to do anything, or you might relax a bit too much and fall asleep.

Even if you don't come to any conclusions or gain any perspective in your time off from a problem, chances are you will return to it both reenergized and ready to expend more effort. In 1927, Gestalt psychologist Bluma Zeigarnik noticed a funny thing: waiters in a Vienna restaurant could remember only orders that were in progress. As soon as the order was sent out and complete, they seemed to wipe it from memory. Zeigarnik then did what any good psychologist would do: she went back to the lab and designed a study. A group of adults and children was given anywhere between eighteen and twenty-two tasks to perform (both physical ones, like making clay figures, and mental ones, like solving puzzles), but half of those tasks were interrupted so that they couldn't be completed. At the end, the subjects remembered the interrupted tasks far better than the completed ones—over two times better, in fact.

Zeigarnik ascribed the finding to a state of tension, akin to a cliff-hanger ending. Your mind *wants* to know what comes next. It wants to finish. It wants to keep working—and it will keep working even if you tell it to stop. All through those other tasks, it will subconsciously be remembering the ones it never got to complete. It's the same Need for Closure that we've encountered before, a desire of our minds to end states of uncertainty and resolve unfinished business. This need motivates us to work harder, to work better, and to work to completion. And a motivated mind, as we already know, is a far more powerful mind.

Distancing Through Actual Distance

And what if, like Watson, you simply can't fathom doing something that would enable you to think of something else, even if you have all of these suggestions to choose from? Luckily, distance isn't limited to a change in activity (though that does happen to be one of the easier routes). Another way to cue psychological distance is to acquire literal distance. To physically move to another point. For Watson, that would be the equivalent of getting up and walking out of Baker Street instead of sitting there looking at his flatmate. Holmes may be able to change location mentally, but an actual physical change may help the lesser willed—and could even aid the great detective himself when imaginative inspiration is not otherwise forthcoming.

In *The Valley of Fear*, Holmes proposes to return in the evening to the scene of the crime under investigation, leaving the hotel where he has been doing most of his thinking.

"An evening alone!" Watson exclaims. Surely, that would be more morbid than anything else. Nonsense, Holmes counters. It could actually be quite illustrative. "I propose to go up there presently. I have arranged it with the estimable Ames, who is by no means whole-hearted about Barker. I shall sit in that room and see if its atmosphere brings me inspiration. I'm a believer in the *genius loci*. You smile, Friend Watson. Well, we shall see." And with that, Holmes is off to the study.

And does he find inspiration? He does. The next morning he is ready with his solution to the mystery. How is that possible? Could the genius loci have really brought the inspiration that Holmes had hoped?

Indeed it could have. Location affects thought in the most direct way possible—in fact, it even affects us physically. It all goes back to one of the most famous experiments in psychology: Pavlov's dogs. Ivan Pavlov wanted to show that a physical cue (in this case it was a sound, but it can just as easily be something visual or a smell or a general location) could eventually elicit the same response as an actual reward. So, he would ring a bell and then present his dogs with food. At the sight of the food, the dogs would—naturally—salivate. But soon enough, they began to salivate at the bell itself, before any sight or smell of food was present. The bell triggered the anticipation of food and with it, a physical reaction.

We now know that this type of learned association goes far beyond dogs and bells and meat. Humans tend to build such patterns as a matter of course, eventually leading innocuous things like bells to trigger predictable reactions in our brains. When you enter a doctor's office, for example, the smell alone may be enough to trigger butterflies—not because you know there will be something painful (you might be coming in to drop off some forms, for all that) but because you have learned to associate that environment with the anxiety of a medical visit.

The power of learned associations is ubiquitous. We tend, for instance, to remember material better in the location where we first learned it. Students who take tests in the room where they did their studying tend to do better than if they take those same tests in a new environment. And the opposite is true: if a particular location is tied to frustration or boredom or distraction, it doesn't make for a good study choice.

At every level, physical and neural, locations get linked to memories. Places tend to get associated with the type of activity that occurs there, and the pattern can be remarkably difficult to break. Watching television in bed, for instance, may make it difficult to get to sleep (unless, that is, you go to sleep while watching TV). Sitting at the same desk all day may make it difficult to unstick yourself if your mind gets stuck.

The tie between location and thought explains why so many people can't work from home and need to go to a specified office. At home, they are not used to working, and they find themselves being distracted by the same types of things that they would normally do around the house. Those neural associations are not the ones that would be conducive to getting things—work-related things, that is—done. The memory traces simply aren't there, and the ones that are there aren't the ones you want to activate. It also illustrates why walking might be so effective. It's much more difficult to fall into a counterproductive thought pattern if your scenery is changing all the time.

Location affects thought. A change in location cues us, so to speak, to think differently. It renders our ingrained associations irrelevant and, in so doing frees us to form new ones, to explore ways of thinking and paths of thought that we hadn't previously considered. Whereas our imagination may be stymied by our usual locations, it is set loose when we separate it from learned constraints. We have no memories, no neural links that kick in to tie us down. And in that lies the secret link between imagination and physical distance. The most important thing that a change in physical perspective can do is to prompt a change in mental perspective. Even Holmes, who unlike Watson doesn't need to be led by the hand and forcibly removed from Baker Street in order to profit from some mental distance, benefits from this property.

Let's return once more to Holmes's strange request in *The Valley of Fear* to spend his night alone in the room where a murder has taken place. In light of the link between location, memory, and imaginative distance, his belief in the genius loci no longer seems nearly as strange. Holmes doesn't actually think that he can re-create events by being in the room where they took place; instead, he banks on doing precisely what we've just discussed. He wants to trigger a change of perspective by a literal change of location, in this case a very specific location and a very specific perspective, that of the people involved in the crime at hand. In doing so, he frees up his imagination to take not the path of his own experiences, memories, and connections but that of the people involved in the events themselves. What associations might the room have triggered for them? What might it have inspired?

Holmes realizes both the necessity of getting into the mindset of the actors involved in the drama and the immediate difficulty of doing so, with all of the elements that could at any point go wrong. And what better way to push all distracting information to the side and focus on the most basic particulars, in a way that is most likely to recall that of the original actors, than to request a solitary evening in the room of the crime? Of course, Holmes still needs all of his observational and imaginative skills once he is there—but he now has access to the tableau and elements that presented themselves to whomever was present at the original scene of the crime. And from there he can proceed on a much more sure footing.

Indeed, it is in that room that he first notices a single dumbbell, surmising at once that the missing member of the pair must have somehow been involved in the unfolding events, and from that room that he deduces the most likely location of the dumbbell's pair: out the only window from which it could reasonably have been dropped. And when he emerges from the study, he has changed his mind from his original conjectures as to the proper course of events. While there, he was better able to get into the mindset of the actors in question and in so doing clarify the elements that had previously been hazy.

And in that sense, Sherlock Holmes invokes the same contextual memory principle as we just explored, using context to cue perspective taking and imagination. Given this specific room, at this specific time of day, what would someone who was committing or had just committed the crime in question be most likely to do or think?

Absent the physical change and distance, however, even Holmes may have found his imagination faltering, as indeed he did prior to that evening, in failing to conceive of the actual course of events as one of the possibilities. We are not often trained to look at the world from another's point of view in a more basic, broad fashion that transcends simple interaction. How might someone else interpret a situation differently from us? How might he act given a specific set of circumstances? What might he think given certain inputs? These are not questions that we often find ourselves asking.

Indeed, so poorly trained are we at *actually* taking someone else's point of view that when we are explicitly requested to do so, we still proceed from an egocentric place. In one series of studies, researchers found that people adopt the perspective of others by simply adjusting from their own. It's a question of degree rather than type: we tend to begin with our own view as an anchoring point, and then adjust slightly in one direction instead of altering the view altogether. Moreover, once we reach an estimate that sounds satisfactory to us, we stop thinking and consider the problem resolved. We've successfully captured the required point of view. That tendency is known as satisficing, a blend of sufficing and satisfying: a response bias that errs on the egocentric side of plausible answers to a given question. As soon as we find an answer that satisfies, we stop looking, whether or not the answer is ideal or even remotely accurate. (In a recent study of online behavior, for instance, individuals were profoundly influenced by existing personal preferences in their evaluations of websites—and they used those preferences as an anchor to reduce the number of sites they considered and to terminate their online search. As a result, they returned often to already known sites, instead of taking the time to evaluate potential new sources of information, and they chose to focus on search engine summaries instead of using actual site visits to make their decisions.) The tendency toward an egocentric bias in satisficing is especially strong when a plausible answer is presented early on in the search process. We then tend to consider our task complete, even if it's far from being so.

A change in perspective, in physical location, quite simply forces mindfulness. It forces us to reconsider the world, to look at things from a different angle. And sometimes that change in perspective can be the spark that makes a difficult decision manageable, or that engenders creativity where none existed before.

Consider a famous problem-solving experiment, originally designed by Norman Maier in 1931. A participant was placed in a room where two strings were hanging from the ceiling. The participant's job was to tie the two strings together. However, it was impossible to reach one string while holding the other. Several items were also available in the room, such as a pole, an extension cord, and a pair of pliers. What would you have done?

Most participants struggled with the pole and the extension cord, trying their best to reach the end while holding on to the other string. It was tricky business.

The most elegant solution? Tie the pliers to the bottom of one string, then use it as a pendulum and catch it as it floats toward you while you hold the other string. Simple, insightful, quick.

But very few people could visualize the change in object use (here, imagining the pliers as something other than pliers, a weight that could be tied to a string) while embroiled in the task. Those that did did one thing differently: they stepped back. They looked at it from a literal distance. They saw the whole and then tried to envision how they could make the details work. Some did this naturally; some had to be prompted by the experimenter, who seemingly by accident brushed one of the strings to induce a swinging motion (that action was enough to get participants to spontaneously think of the pliers solution). But none did it without a shift, however slight, of point of view, or, to speak in Trope's terms, a move from the concrete (pliers) to the abstract (pendulum weight), from those puzzle pieces to the overall puzzle. Never underestimate how powerful a cue physical perspective can be. As Holmes puts it in "The Problem of Thor Bridge," "When once your point of view is changed, the very thing which was so damning becomes a clue to the truth."

Distancing Through Mental Techniques

Let's return for a moment to a scene that we've visited once before, in *The Hound of the Baskervilles*. After Dr. Mortimer's initial visit, Dr. Watson leaves Baker Street to go to his club. Holmes, however, remains seated in his armchair, which is where Watson finds him when he returns to the flat around nine o'clock in the evening. Has Holmes been a fixture there all day? Watson inquires. "On the contrary," responds Holmes. "I have been to Devonshire." Watson doesn't miss a beat. "In spirit?" he asks. "Exactly," responds the detective.

What is it, exactly, that Holmes does as he sits in his chair, his mind far away from the physicality of the moment? What happens in his brain—and why is it such an effective tool of the imagination, such an important element of his thought process that he hardly ever abandons it? Holmes's mental journeying goes by many names, but most commonly it is called meditation.

When I say *meditation*, the images invoked for most people will include monks or yogis or some other spiritual-sounding monikers. But that is only a tiny portion of what the word means. Holmes is neither monk nor yoga practitioner, but he understands what meditation, in its essence, actually is—a simple mental exercise to clear your mind. Meditation is nothing more than the quiet distance that you need for integrative, imaginative, observant, and mindful thought. It is the ability to create distance, in both time and space, between you and all of the problems you are trying to tackle, in your mind alone. It doesn't even have to be, as people often assume, a way of experiencing nothing; directed meditation can take you toward some specific goal or destination (like Devonshire), as long as your mind is clear of every other distraction—or, to be more precise, as long as your mind clears itself of every distraction and continues to do so as the distractions continue to arise (as they inevitably will).

In 2011, researchers from the University of Wisconsin studied a group of people who were not in the habit of meditating and instructed them in the following manner: relax with your eyes closed and focus on the flow of your breath at the tip of your nose; if a random thought arises, acknowledge the thought and then simply let it go by gently bringing your attention back to the flow of your breath. For fifteen minutes, the participants attempted to follow these guidelines. Then they were broken up into two groups: one group had the option of receiving nine thirty-minute sessions of meditation instruction over the course of five weeks, and the other group had that option at the conclusion of the experiment, but not before. At the end of the five weeks, everyone completed the earlier thought assignment a second time.

During each session, the researchers measured participants' electro-encephalographic (EEG) activity—a recording of electrical activity along the scalp—and what they found presents a tantalizing picture. Even such a short training period—participants averaged between five and sixteen minutes of training and practice a day—can cause changes at the neural level. The researchers were particularly interested in frontal EEG asymmetry, toward a pattern that has been associated with positive emotions (and that had been shown to follow seventy or more hours of training in mindfulness meditation techniques). While prior to training the two groups showed no differences, by the end of the study, those who had received additional training showed a leftward shift in asymmetry, which means a move toward a pattern that has been associated with positive and approach-oriented emotional states—such states as have been linked repeatedly to increased creativity and imaginative capacity.

What does that mean? First, unlike past studies of meditation that asked for a very real input of time and energy, this experiment did not require extensive resource commitment, and yet it still showed striking neural results. Moreover, the training provided was extremely flexible: people could choose when they would want to receive instruction and when they would want to practice. And, perhaps more important, participants reported a spike in spontaneous passive practice, when, without a conscious decision to meditate, they found themselves in unrelated situations thinking along the lines of the instructions they had been provided.

True, it is only one study. But there's more to the brain story than that. Earlier work suggests that meditation training can affect the default network—that diffuse attentional network that we've already talked about, that facilitates creative insights and allows our brains to work on remote connections while we're doing something else entirely. Individuals who meditate regularly show increased resting-state functional connectivity in the network compared to nonmeditators. What's more, in one study of meditation's effects over a period of eight weeks, researchers found changes in gray-matter density in a group of meditation-naive participants (that is, they hadn't practiced meditation before the beginning of the study) as compared to a control group. There were increases in concentration in the left hippocampus, the posterior cingulate cortex (PCC), the tempero-parietal junction (TPJ), and the cerebellum—areas involved in learning and memory, emotion regulation, self-referential processing, and perspective taking. Together, the hippocampus, PCC, and TPJ form a neural network that supports both self-projection—including thinking about the hypothetical future—and perspective taking, or conceiving others' point of view—in other words, precisely the type of distancing that we've been discussing.

Meditation is a way of thinking. A habit of distance that has the fortunate consequence of being self-reinforcing. One tool in the arsenal of mental techniques that can help you create the right frame of mind to attain the distance necessary for mindful, imaginative thought. It is far more attainable, and far more widely applicable, than the connotations of the word might have you believe.

Consider the case of someone like Ray Dalio. Almost every morning, Dalio meditates. Sometimes he does it before work. Sometimes in his office, right at his desk: he leans back, closes his eyes, clasps his hands in a simple grip. Nothing more is necessary. "It's just a mental exercise in which you are clearing your mind," he once told the *New Yorker* in an interview.

Dalio isn't the person that comes to mind most readily when you think of practitioners of meditation. He isn't a monk or a yoga fanatic or a hippie New Ager, and he isn't doing this just for the interest in participating in a psych study. He happens to be the founder of the world's biggest hedge fund, Bridgewater Associates, someone who has little time to waste and many ways to spend the time he does have. And yet he chooses, actively, to devote a portion of each day to mediation, in its broadest, most classic sense.

When Dalio meditates, he clears his mind. He prepares it for the day by relaxing and trying to keep all of the thoughts that will proceed to bother him for the next however many hours at bay. Yes, it may seem like a waste to spend any time at all doing, well, nothing that looks productive. But spending those minutes in the space of his mind will actually make Dalio more productive, more flexible, more imaginative, and more insightful. In short, it will help him be a better decision maker.

But is it for everyone? Meditation, that mental space, is not nothing; it requires real energy and concentration (hence the easier route of physical distance). While someone like Holmes or Dalio may well be able to dive right into blankness to great effect, I'm willing to bet that Watson would struggle. With nothing else to occupy his mind, his breathing alone would likely not be enough to keep all those thoughts in check. It's far easier to distance yourself with physical cues than it is to have to rely on your mind alone.

Luckily, as I mentioned in passing, meditation need not be blank. In meditation, we can indeed be focusing on something as difficult to capture as breath or emotion or the sensations of the body to the exclusion of everything else. But we can also use what's known as visualization: a focus on a specific mental image that will replace that blankness with something more tangible and accessible. Go back for a moment to *The Hound of the Baskervilles*, where we left Holmes floating above the Devonshire moors. That, too, is meditation—and it wasn't at all aimless or blank or devoid of mental imagery. It requires the same focus as any meditation, but is in some ways more approachable. You have a concrete plan, something with which to occupy your mind and keep intrusive thoughts at bay, something on which you can focus your energy that is more vibrant and multidimensional than the rise and fall of your breath. And what's more, you can focus on attaining the distance that Trope would call hypotheticality, to begin considering the ifs and what-ifs.

Try this exercise. Close your eyes (well, close them once you finish reading the instructions). Think of a specific situation where you felt angry or hostile, your most recent fight with a close friend or significant other, for instance. Do you have a moment in mind? Recall it as closely as you can, as if you were going through it again. Once you're done, tell me how you feel. And tell me as far as you can what went wrong. Who was to blame? Why? Do you think it's something that can be fixed?

Close your eyes again. Picture the same situation. Only now, I want you to imagine that it is happening to two people who are not you. You are just a small fly on the wall, looking down at the scene and taking note of it. You are free to buzz around and observe from all angles and no one will see you. Once again, as soon as you finish, tell me how you feel. And then respond to the same questions as before.

You've just completed a classic exercise in mental distancing through visualization. It's a process of picturing something vividly but from a distance, and so, from a perspective that is inherently different from the actual one you have stored in your memory. From scenario one to scenario two, you have gone from a concrete to an abstract mindset; you've likely become calmer emotionally, seen things that you missed the first time around, and you may have even come away with a slightly modified memory of what happened. In fact, you may have even become wiser and better at solving problems *overall*, unrelated to the scenario in question. (And you will have also been practicing a form of meditation. Sneaky, isn't it?)

Psychologist Ethan Kross has demonstrated that such mental distancing (the above scenario was actually taken from one of his studies) is not just good for emotional regulation. It can also enhance your wisdom, both in terms of dialectism (i.e., being cognizant of change and contradictions in the world) and intellectual humility (i.e., knowing your own limitations), and make you better able to solve problems and make choices. When you distance yourself, you begin to process things more broadly, see connections that you couldn't see from a closer vantage point. In other words, being wiser also means being more imaginative. It might not lead to a eureka moment, but it will lead to insight. You think *as if* you had actually changed your location, while you remain seated in your armchair.

Jacob Rabinow, an electrical engineer, was one of the most talented and prolific inventors of the twentieth century. Among his 230 U.S. patents is the automatic letter-sorting machine that the postal service still uses to sort the mail, a magnetic memory device that served as a precursor to the hard disk drive, and the straight-arm phonograph. One of the tricks that helped sustain his remarkable creativity and productivity? None other than visualization. As he once told psychologist Mihaly Csikszentmi-halyi, whenever a task proves difficult or takes time or doesn't have an ob-vious answer, "I pretend I'm in jail. If I'm in jail, time is of no consequence. In other words, if it takes a week to cut this, it'll take a week. What else have I got to do? I'm going to be here for twenty years. See? This is a kind of mental trick. Otherwise you say, 'My God, it's not working,' and then you make mistakes. My way, you say time is of absolutely no conse-quence." Visualization helped Rabinow to shift his mindset to one where he was able to tackle things that would otherwise overwhelm him, pro-viding the requisite imaginative space for such problem solving to occur.

The technique is widespread. Athletes often visualize certain ele-ments of a game or move before they actually perform them, acting them out in their minds before they do so in reality: a tennis player envisions a serve before the ball has left his hand; a golfer sees the path of the ball before he lifts his club. Cognitive behavioral therapists use the technique to help people who suffer from phobias or other conditions to relax and be able to experience situations without actually experiencing them. Psy-chologist Martin Seligman urges that it might even be the single most important tool toward fostering a more imaginative, intuitive mindset. He goes as far as to suggest that by repeated, simulated visual enactment, "intuition may be teachable virtually and on a massive scale." How's that for endorsement.

It is all about learning to create distance with the mind by actually picturing a world as if you were seeing and experiencing it for real. As the philosopher Ludwig Wittgenstein once put it, "To repeat: don't think, but look!" That is the essence of visualization: learning to look internally, to create scenarios and alternatives in your mind, to play out nonrealities as if they were real. It helps you see beyond the obvious, to not make the mistakes of a Lestrade or a Gregson by playing through only the scenario that is in front of you, or the only one you want to see. It forces imagina-tion because it necessitates the use of imagination.

It's easier than you might think. In fact, all it is really is what we do naturally when we try to recall a memory. It even uses the same neural network—the MPFC, lateral temporal cortex, medial and lateral parietal lobes, and the medial temporal lobe (home of the hippocampus). Except, instead of recalling a memory exactly, we shuffle around details from experience to create something that never actually occurred, be it a not-yet-extant future or a counterfactual past. We test it in our minds instead of having to experience it in reality. And by so doing, we attain the very same thing we do by way of physical distance: we separate ourselves from the situation we are trying to analyze.

It is all meditation of one form or another. When we saw Holmes in *The Valley of Fear*, he asked for a physical change in location, an actual prompt for his mind from the external world. But the same effect can be accomplished without having to go anywhere—from behind your desk, if you're Dalio, or your armchair, if you're Holmes, or wherever else you might find yourself. All you have to do is be able to free up the necessary space in your mind. Let it be the blank canvas. And then the whole imaginative world can be your palette.

Sustaining Your Imagination: The Importance of Curiosity and Play

Once upon a time, Sherlock Holmes urged us to maintain a crisp and clean brain attic: out with the useless junk, in with meticulously organized boxes that are uncluttered by useless paraphernalia. But it's not quite that simple. Why on earth, for instance, did Holmes, in "The Lion's Mane," know about an obscure species of jellyfish in one warm corner of the ocean? Impossible to explain it by virtue of the stark criteria he imposes early on. As with most things, it is safe to assume that Holmes was exaggerating for effect. Uncluttered, yes, but not stark. An attic that contained only the bare essentials for your professional success would be a sad little attic indeed. It would have hardly any material to work with, and it would be practically incapable of any great insight or imagination.

How did the jellyfish make its way into Holmes's pristine palace? It's simple. At some point Holmes must have gotten curious. Just like he got curious about the Motets. Just like he gets curious about art long enough to try to convince Scotland Yard that his nemesis, Professor Moriarty, can't possibly be up to any good. Just as he says to Inspector MacDonald in *The Valley of Fear*, when the inspector indignantly refuses Holmes's offer of reading a book on the history of Manor House, "Breadth of view, my dear Mr. Mac, is one of the essentials of our profession. The interplay of ideas and the oblique uses of knowledge are often of extraordinary interest." Time and time again, Holmes gets curious, and his curiosity leads him to find out more. And that "more" is then tucked away in some obscure (but labeled!) box in his attic.

For that is basically what Holmes is telling us. Your attic has levels of storage.

There is a difference between active and passive knowledge, those boxes that you need to access regularly and as a matter of course and those that you may need to reach one day but don't necessarily look to on a regular basis. Holmes isn't asking that we stop being curious, that we stop acquiring those jellyfish. No. He asks that we keep the active knowledge clean and clear—and that we store the passive knowledge cleanly and clearly, in properly labeled boxes and bins, in the right folders and the right drawers.

It's not that we should all of a sudden go against his earlier admonition and take up our precious mental real estate with junk. Not at all. Only, we don't always know when something that may at first glance appear to be junklike is not junk at all but an important addition to our mental arsenal. So, we must tuck those items away securely in case of future use. We don't even need to store the full item; just a trace of what it was, a reminder that will allow us to find it again—just as Holmes looks up the jellyfish particulars in an old book rather than knowing them as a matter of course. All he needs to do is remember that the book and the reference exist.

An organized attic is not a static attic. Imagination allows you to make more out of your mind space than you otherwise could. And the truth is you never quite know what element will be of most use and when it might end up being more useful than you ever thought possible.

Here, then, is Holmes's all-important caveat: the most surprising of articles can end up being useful in the most surprising of ways. You must open your mind to new inputs, however unrelated they may seem.

And that is where your general mindset comes in. Is there a standing openness to inputs no matter how strange or unnecessary they might seem, as opposed to a tendency to dismiss anything that is potentially distracting? Is that open-minded stance your habitual approach, the way that you train yourself to think and to look at the world?

With practice, we might become better at sensing what may or may not prove useful, what to store away for future reference and what to throw out for the time being. Something that at first glance may seem like simple intuition is actually far more—a knowledge that is actually based on countless hours of practice, of training yourself to be open, to integrate experiences in your mind until you become familiar with the patterns and directions those experiences tend to take.

Remember those remote-association experiments, where you had to find a word that could complete all three members of a set? In a way, that encapsulates most of life: a series of remote associations that you won't see unless you take the time to stop, to imagine, and to consider. If your mindset is one that is scared of creativity, scared to go against prevailing customs and mores, it will only hold you back. If you fear creativity, even subconsciously, you will have more difficulty being creative. You will never be like Holmes, try as you may. Never forget that Holmes was a renegade—and a renegade that was as far from a computer as it gets. And that is what makes his approach so powerful.

Holmes gets to the very heart of the matter in *The Valley of Fear*, when he admonishes Watson that "there should be no combination of events for which the wit of man cannot conceive an explanation. Simply as a mental exercise, without any assertion that it is true, let me indicate a possible line of thought. It is, I admit, mere imagination; but how often is imagination the mother of truth?"

SHERLOCK HOLMES FURTHER READING

"Here is a young man who learns suddenly . . ." "Not until I have been to Blackheath." from *The Casebook of Sherlock Holmes*, "The Adventure of the Norwood Builder," p. 829.

"You will rise high in your profession." from *His Last Bow*, "The Adventure of Wisteria Lodge," p. 1231.

"One of the most remarkable characteristics of Sherlock Holmes was

his power of throwing his brain out of action . . ." from *His Last Bow,* "The Adventure of the Bruce-Partington Plans," p. 297.

"It is quite a three-pipe problem . . ." from *The Adventures of Sherlock Holmes,* "The Red-Headed League," p. 50.

"I have been to Devonshire." from *The Hound of the Baskervilles,* chapter 3: The Problem, p. 22.

"I'm a believer in the genius loci." *"Breadth of view, my dear Mr. Mac, is one of the essentials of our profession."* from chapter 6: A Dawning Light, p. 51; chapter 7: The Solution, p. 62 *The Valley of Fear.*

PART THREE

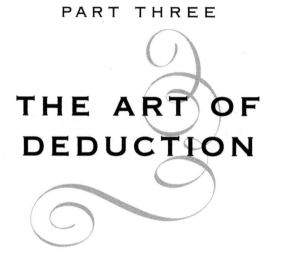

THE ART OF DEDUCTION

CHAPTER FIVE

Navigating the Brain Attic:
Deduction from the Facts

I magine you are Holmes, and I, Maria, a potential client. You've spent the last hundred-odd pages being presented with information, much as you would if you were to observe me in your sitting room for some time. Take a minute to think, to consider what you may know about me as a person. What can you infer based on what I've written?

I won't go down the list of all possible answers, but here's one to make you pause: the first time I ever heard the name Sherlock Holmes was in Russian. Those stories my dad read by the fire? Russian translations, not English originals. You see, we had only recently come to the United States, and when he read to us, it was in the language that my family uses to this day with one another at home. Alexandre Dumas, Sir H. Rider Haggard, Jerome K. Jerome, Sir Arthur Conan Doyle: all men whose voices I first heard in Russian.

What does this have to do with anything? Simply this: Holmes would have known without my having to tell him. He would have made a simple deduction based on the available facts, infused with just a bit of that imaginative quality we spoke about in the last chapter. And he would have realized that I couldn't have possibly had my first encounter with his methods in any language but Russian. Don't believe me? All of the elements are there, I promise. And by the end of this chapter, you, too, should be in a position to follow Holmes in putting them together into the only explanation that would suit all of the available facts. As the detective says over and over, when all avenues are exhausted, whatever remains, however improbable, must be the truth.

.   .   .

And so we turn finally to that most flashy of steps: deduction. The grand finale. The fireworks at the end of a hard day's work. The moment when you can finally complete your thought process and come to your conclusion, make your decision, do whatever it was that you had set out to do. Everything has been gathered and analyzed. All that remains is to see what it all means and what that meaning implies for you, to draw the implications out to their logical conclusion.

It's the moment when Sherlock Holmes utters that immortal line in "The Crooked Man," *elementary.*

> "I have the advantage of knowing your habits, my dear Watson," said he. "When your round is a short one you walk, and when it is a long one you use a hansom. As I perceive that your boots, although used, are by no means dirty, I cannot doubt that you are at present busy enough to justify the hansom."
>
> "Excellent!" I cried.
>
> "Elementary," said he. "It is one of those instances where the reasoner can produce an effect which seems remarkable to his neighbour, because the latter has missed the one little point which is the basis of the deduction."

What does deduction actually entail? Deduction is that final navigation of your brain attic, the moment when you put together all of the elements that came before in a single, cohesive whole that makes sense of the full picture, the attic yielding in orderly fashion what it has gathered so methodically. What Holmes means by deduction and what formal logic means by deduction are not one and the same. In the purely logical sense, deduction is the arrival at a specific instance from a general principle. Perhaps the most famous example:

> All men are mortal.
> Socrates is a man.
> Socrates is mortal.

But for Holmes, this is but one possible way to reach the conclusion. His deduction includes multiple ways of reasoning—as long as you proceed from fact and reach a statement that must necessarily be true, to the exclusion of other alternatives.[3]

Whether it's solving a crime, making a decision, or coming to some personal determination, the process remains essentially the same. You take all of your observations—those attic contents that you've decided to store and integrate into your existing attic structure and that you've already mulled over and reconfigured in your imagination—you put them in order, starting from the beginning and leaving nothing out, and you see what possible answer remains that will both incorporate all of them and answer your initial question. Or, to put it in Holmesian terms, you lay out your chain of reasoning and test possibilities until whatever remains (improbability aside) is the truth: "That process starts upon the supposition that when you have eliminated all which is impossible, then whatever remains, however improbable, must be the truth," he tells us. "It may well be that several explanations remain, in which case one tries test after test until one or other of them has a convincing amount of support."

That, in essence, is deduction, or what Holmes calls "systematized common sense." But the common sense is not as common, or as straightforward, as one might hope. Whenever Watson himself tries to emulate Holmes, he often finds himself in error. And it's only natural. Even if we've been accurate up to this point, we have to push back one more time lest System Watson leads us astray at the eleventh hour.

Why is deduction far more difficult than it appears? Why is it that Watson so often falters when he tries to follow in his companion's footsteps. What gets in the way of our final reasoning? Why is it so often so difficult to think clearly, even when we have everything we need to do so? And how can we circumvent those difficulties so that, unlike Watson, who is stuck to repeat his mistakes over and over, we can use System Holmes to help us out of the quagmire and deduce properly?

---

[3] Indeed, some of his deduction would, in logic's terms, be more properly called induction or abduction. All references to deduction or deductive reasoning use it in the Holmesian sense, and not the formal logic sense.

The Difficulty of Proper Deduction:
Our Inner Storyteller at the Wheel

A trio of notorious robbers sets its sights on Abbey Grange, the residence of Sir Eustace Brackenstall, one of the richest men in Kent. One night, when all are presumed to be sleeping, the three men make their way through the dining room window, preparing to ransack the wealthy residence much as they did a nearby estate a fortnight prior. Their plan, however, is foiled when Lady Brackenstall enters the room. Quickly, they hit her over the head and tie her to one of the dining room chairs. All would seem to be well, were it not for Sir Brackenstall, who comes in to investigate the strange noises. He is not so lucky as his wife: he is knocked over the head with a poker and he collapses, dead, onto the floor. The robbers hastily clear the sideboard of its silver but, too agitated by the murder to do much else, exit thereafter. But first they open a bottle of wine to calm their nerves.

Or so it would seem, according to the testimony of the only living witness, Lady Brackenstall. But in "The Adventure of the Abbey Grange," few things are what they appear to be.

The story seems sound enough. The lady's explanation is confirmed by her maid, Theresa, and all signs point to events unfolding much in the manner she has described. And yet, something doesn't feel right to Sherlock Holmes. "Every instinct that I possess cries out against it," he tells Watson. "It's wrong—it's all wrong—I'll swear that it's wrong." He begins enumerating the possible flaws, and as he does so, details that seem entirely plausible, when taken one by one, now together begin to cast doubt on the likelihood of the story. It is not, however, until he comes to the wineglasses that Holmes knows for sure he is correct. "And now, on the top of this, comes the incident of the wineglasses," he says to his companion.

> "Can you see them in your mind's eye?"
> "I see them clearly."
> "We are told that three men drank from them. Does that strike you as likely?"

"Why not? There was wine in each glass."

"Exactly, but there was beeswing only in one glass. You must have noticed that fact. What does that suggest to your mind?"

"That last glass filled would be most likely to contain beeswing."

"Not at all. The bottle was full of it, and it is inconceivable that the first two glasses were clear and the third heaving charged with it. There are two possible explanations, and only two. One is that after the second glass was filled the bottle was violently agitated, and so the third glass received the beeswing. That does not appear probable. No, no, I am sure that I am right."

"What, then, do you suppose?"

"That only two glasses were used, and that the dregs of both were poured into a third glass, so as to give the false impression that three people had been there."

What does Watson know about the physics of wine? Not much, I venture to guess, but when Holmes asks him about the beeswing, he at once comes up with a ready answer: it must have been the last glass to be poured. The reason seems sensible enough, and yet comes from nowhere. I'd bet that Watson hadn't even given it so much as a second thought until Holmes prompted him to do so. But when asked, he is only too happy to create an explanation that makes sense. Watson doesn't even realize that he has done it, and were Holmes not to stop him for a moment, he would likely hold it as future fact, as further proof of the veracity of the original story rather than as a potential hole in the story's fabric.

Absent Holmes, the Watson storytelling approach is the natural, instinctive one. And absent Holmes's insistence, it is incredibly difficult to resist our desire to form narratives, to tell stories even if they may not be altogether correct, or correct at all. We like simplicity. We like concrete reasons. We like causes. We like things that make intuitive sense (even if that sense happens to be wrong).

On the flip side, we dislike any factor that stands in the way of that simplicity and causal concreteness. Uncertainty, chance, randomness, nonlinearity: these elements threaten our ability to explain, and to explain quickly and (seemingly) logically. And so, we do our best to eliminate them at every turn. Just like we decide that the last glass of wine to be poured is also most likely to contain all the beeswing if we see glasses of uneven clarity, we may think, to take one example, that someone has a hot hand in basketball if we see a number of baskets in a row (the hot-hand fallacy). In both cases, we are using too few observations to reach our conclusions. In the case of the glasses, we rely only on that bottle and not on the behavior of other similar bottles under various circumstances. In the case of basketball, we rely only on the short streak (the law of small numbers) and not on the variability inherent in any player's game, which includes long-run streaks. Or, to take another example, we think a coin is more likely to land on heads if it has fallen on tails for a number of times (the gambler's fallacy), forgetting that short sequences don't necessarily have to have the fifty-fifty distribution that would appear in the long term.

Whether we're explaining why something has happened or concluding as to the likely cause of an event, our intuition often fails us because we prefer things to be much more controllable, predictable, and causally determined than they are in reality.

From these preferences stem the errors in thinking that we make without so much as a second thought. We tend to deduce as we shouldn't, arguing, as Holmes would put it, ahead of the data—and often in spite of the data. When things just "make sense" it is incredibly difficult to see them any other way.

W.J. was a World War II veteran. He was gregarious, charming, and witty. He also happened to suffer from a form of epilepsy so incapacitating that, in 1960, he elected to have a drastic form of brain surgery. The connecting fabric between the left and right hemispheres of the brain that allows the two halves to communicate—his corpus collosum—would be severed. In the past, this form of treatment had been shown to have a dramatic effect on the incidence of seizures. Patients who had been unable to function could all of a sudden lead seizure-free lives. But did such a dramatic change to the brain's natural connectivity come at a cost?

At the time of W.J.'s surgery, no one really knew the answer. But Roger Sperry, a neuroscientist at Caltech who would go on to win a Nobel Prize in medicine for his work on hemispheric connectictivity, suspected that

it might. In animals, at least, a severing of the corpus collosum meant
that the hemispheres became unable to communicate. What happened in
one hemisphere was now a complete mystery to the other. Could this effective isolation occur in humans as well?

The pervasive wisdom was an emphatic no. Our human brains were
not animal brains. They were far more complicated, far too smart, far too
evolved, really. And what better proof than all of the high-functioning
patients who had undergone the surgery. This was no frontal lobotomy.
These patients emerged with IQ intact and reasoning abilities aplenty.
Their memory seemed unaffected. Their language abilities were normal.

The resounding wisdom seemed intuitive and accurate. Except, of
course, it was resoundingly wrong. No one had ever figured out a way to
test it scientifically: it was a Watson just-so story that made sense, founded
on the same absence of verified factual underpinnings. Until, that is, the
scientific equivalent of Holmes arrived at the scene: Michael Gazzaniga,
a young neuroscientist in Sperry's lab. Gazzaniga found a way to test
Sperry's theory—that a severed corpus collosum rendered the brain hemispheres unable to communicate—with the use of a tachistoscope, a device
that could present visual stimuli for specific periods of time, and, crucially, could do this to the right side or the left side of each eye separately.
(This lateral presentation meant that any information would go to only
one of the two hemispheres.)

When Gazzaniga tested W.J. after the surgery, the results were striking. The same man who had sailed through his tests weeks earlier could
no longer describe a single object that was presented to his left visual
field. When Gazzaniga flashed an image of a spoon to the right field, W.J.
named it easily, but when the same picture was presented to the left, the
patient seemed to have, in essence, gone blind. His eyes were fully functional, but he could neither verbalize nor recall having seen a single
thing.

What was going on? W.J. was Gazzaniga's patient zero, the first in a
long line of initials who all pointed in one direction: the two halves of our
brains are not created equal. One half is responsible for processing visual
inputs—it's the one with the little window to the outside world, if you
recall the Shel Silverstein image—but the other half is responsible for

verbalizing what it knows—it's the one with the staircase to the rest of the house. When the two halves have been split apart, the bridge that connects the two no longer exists. Any information available to one side may as well not exist as far as the other is concerned. We have, in effect, two separate mind attics, each with its unique storage, contents, and, to some extent, structure.

And here's where things get really tricky. If you show a picture of, say, a chicken claw to just the left side of the eye (which means the picture will be processed only by the right hemisphere of the brain—the visual one, with the window) and one of a snowy driveway to just the right side of the eye (which means it will be processed only by the left hemisphere—the one with the communicating staircase), and then ask the individual to point at an image most closely related to what he's seen, the two hands don't agree: the right hand (tied to the left input) will point to a shovel, while the left hand (tied to the right input) will point to a chicken. Ask the person why he's pointing to two objects, and instead of being confused he'll at once create an entirely plausible explanation: you need a shovel to clean out the chicken coop. His mind has created an entire story, a narrative that will make plausible sense of his hands' discrepancy, when in reality it all goes back to those silent images.

Gazzaniga calls the left hemisphere our left-brain interpreter, driven to seek causes and explanations—even for things that may not have them, or at least not readily available to our minds—in a natural and instinctive fashion. But while the interpreter makes perfect sense, he is more often than not flat-out wrong, the Watson of the wineglasses taken to an extreme.

Split-brain patients provide some of the best scientific evidence of our proficiency at narrative self-deception, at creating explanations that make sense but are in reality far from the truth. But we don't even need to have our corpus collosum severed to act that way. We do it all the time, as a matter of course. Remember that pendulum study of creativity, where subjects were able to solve the problem after the experimenter had casually set one of the two cords in motion? When subjects were then asked where their insight had come from, they cited many causes. "It was the only thing left." "I just realized the cord would swing if I fastened a weight to it." "I thought of the situation of swinging across a river." "I had imagery of monkeys swinging from trees."

All plausible enough. None correct. No one mentioned the experimenter's ploy. And even when told about it later, over two-thirds continued to insist that they had not noted it and that it had had no impact at all on their own solutions—although they had reached those solutions, on average, within forty-five seconds of the hint. What's more, even the one-third that admitted the possibility of influence proved susceptible to false explanation. When a decoy cue (twirling the weight on a cord) was presented, which had *no* impact on the solution, they cited that cue, and not the actual one that helped them, as having prompted their behavior.

Our minds form cohesive narratives out of disparate elements all the time. We're not comfortable if something doesn't have a cause, and so our brains determine a cause one way or the other, without asking our permission to do so. When in doubt, our brains take the easiest route, and they do so at every stage of the reasoning process, from forming inferences to generalizations.

W.J. is but a more extreme example of the exact thing that Watson does with the wineglasses. In both instances there is the spontaneous construction of story, and then a firm belief in its veracity, even when it hinges on nothing more than its seeming cohesiveness. That is deductive problem number one.

Even though all of the material is there for the taking, the possibility of ignoring some of it, knowingly or not, is real. Memory is highly imperfect, and highly subject to change and influence. Even our observations themselves, while accurate enough to begin with, may end up affecting our recall and, hence, our deductive reasoning more than we think. We must be careful lest we let something that caught our attention, whether because it is out of all proportion (salience) or because it just happened (recency) or because we've been thinking about something totally unrelated (priming or framing), weigh too heavily in our reasoning and make us forget other details that are crucial for proper deduction. We must also be sure that we answer the same question we posed in the beginning, the one that was informed by our initial goals and motivation, and not one that somehow seems more pertinent or intuitive or easier, now that we've reached the end of the thought process. Why do Lestrade and the rest of the detectives so often persist in wrongful arrests, even when all evidence points to the contrary? Why do they keep pushing their original story, as if failing to note altogether that it is coming apart at the seams? It's simple, really. We don't like to admit our initial intuition to be false and would much rather dismiss the evidence that contradicts it. It is perhaps why wrongful arrests are so sticky even outside the world of Conan Doyle.

The precise mistakes or the names we give them don't matter as much as the broad idea: we often aren't mindful in our deduction, and the temptation to gloss over and jump to the end becomes ever stronger the closer we get to the finish line. Our natural stories are so incredibly compelling that they are tough to ignore or reverse. They get in the way of Holmes's dictate of systematized common sense, of going through *all* alternatives, one by one, sifting the crucial from the incidental, the improbable from the impossible, until we reach the only answer.

As a simple illustration of what I mean, consider the following questions. I want you to write down the first answer that comes to your mind. Ready?

1. A bat and a ball cost $1.10 in total. The bat costs $1.00 more than the ball. How much does the ball cost?
2. If it takes 5 machines 5 minutes to make 5 widgets, how long would it take 100 machines to make 100 widgets?
3. In a lake, there is a patch of lily pads. Every day, the patch doubles in size. If it takes 48 days for the patch to cover the entire lake, how long would it take for the patch to cover half of the lake?

You have just taken Shane Frederick's Cognitive Reflection Test (CRT). If you are like most people, chances are you wrote down at least one of the following: $0.10 for question one; 100 minutes for question two; and 24 days for question three. In each case, you would have been wrong. But you would have been wrong in good company. When the questions were asked of Harvard students, the average score was 1.43 correct (with 57 percent of students getting either zero or one right). At Princeton, a similar story: 1.63 correct, and 45 percent scoring zero or one. And even at MIT, the scores were far from perfect: 2.18 correct on average, with 23 percent, or near to a quarter, of students getting either none or one correct. These "simple" problems are not as straightforward as they may seem at first glance.

The correct answers are $0.05, 5 minutes, and 47 days, respectively. If you take a moment to reflect, you will likely see why—and you'll say to yourself, *Of course, how did I ever miss that?* Simple. Good old System Watson has won out once again. The initial answers are the intuitively appealing ones, the ones that come quickly and naturally to mind if we don't pause to reflect. We let the salience of certain elements (and they were framed to be salient on purpose) draw us away from considering each element fairly and accurately. We use mindless verbatim strategies— repeating an element in the prior answer and not reflecting on the actual best strategy to solve the present problem—instead of mindful ones (in essence, substituting an intuitive question for the more difficult and time-consuming alternative, just because the two happen to seem related). Those second answers require you to suppress System Watson's eager response and let Holmes take a look: to reflect, inhibit your initial intuition, and then edit it accordingly, which is not something that we are overly eager to do, especially when we are tired from all the thinking that came before. It's tough to keep that motivation and mindfulness going from start to finish, and far easier to start conserving our cognitive resources by letting Watson take the helm.

While the CRT may seem far removed from any real problems we might encounter, it happens to be remarkably predictive of our performance in any number of situations where logic and deduction come into play. In fact, this test is often more telling than are measures of cognitive ability, thinking disposition, and executive function. Good performance on these three little questions predicts resistance to a number of common logical fallacies, which, taken together, are considered to predict adherence to the basic structures of rational thought. The CRT even predicts our ability to reason through the type of formal deductive problem—the Socrates one—that we saw earlier in the chapter: if you do poorly on the test, you are more likely to say that if all living things need water and roses need water, it follows that roses are living things.

Jumping to conclusions, telling a selective story instead of a logical one, even with all of the evidence in front of you and well sorted, is common (though avoidable, as you'll see in just a moment). Reasoning through everything up until the last moment, not letting those mundane details bore you, not letting yourself peter out toward the end of the process: that is altogether rare. We need to learn to take pleasure in the lowliest manifestations of reason. To take care that deduction not seem boring, or too simple, after all of the effort that has preceded it. That is a difficult task. In the opening lines of "The Adventure of the Copper Beeches," Holmes reminds us, "To the man who loves art for its own sake, it is frequently in its least important and lowliest manifestations that the keenest pleasure is to be derived. . . . If I claim full justice for my art, it is because it is an impersonal thing—a thing beyond myself. Crime is common. Logic is rare." Why? Logic is *boring*. We think we've already figured it out. In pushing past this preconception lies the challenge.

Learning to Tell the Crucial from the Incidental

So how do you start from the beginning and make sure that your deduction is going along the right track and has not veered fabulously off course before it has even begun?

In "The Crooked Man," Sherlock Holmes describes a new case, the death of Sergeant James Barclay, to Watson. At first glance the facts are strange indeed. Barclay and his wife, Nancy, were heard to be arguing in the morning room. The two were usually affectionate, and so the argument in itself was something of an event. But it became even more striking when the housemaid found the door to the room locked and its occupants unresponsive to her knocks. Add to that a strange name that she heard several times—David—and then the most remarkable fact of all: after the coachman succeeded in entering the room from outside through the open French doors, no key was to be found. The lady was lying insensible on the couch, the gentleman dead, with a jagged cut on the back of his head and his face twisted in horror. And neither one possessed the key that would open the locked door.

How to make sense of these multiple elements? "Having gathered these facts, Watson," Holmes tells the doctor, "I smoked several pipes over them, trying to separate those which were crucial from others which were merely incidental." And that, in one sentence, is the first step toward successful deduction: the separation of those factors that are crucial to your judgment from those that are just incidental, to make sure that only the truly central elements affect your decision.

Consider the following descriptions of two people, Bill and Linda. Each description is followed by a list of occupations and avocations. Your task is to rank the items in the list by the degree that Bill or Linda resembles the typical member of the class.

Bill is thirty-four years old. He is intelligent but unimaginative, compulsive, and generally lifeless. In school he was strong in mathematics but weak in social studies and humanities.
   Bill is a physician who plays poker for a hobby.
   Bill is an architect.
   Bill is an accountant.
   Bill plays jazz for a hobby.
   Bill is a reporter.
   Bill is an accountant who plays jazz for a hobby.
   Bill climbs mountains for a hobby.

Linda is thirty-one years old, single, outspoken, and very bright. She majored in philosophy. As a student, she was deeply concerned with issues of discrimination and social justice, and also participated in antinuclear demonstrations.
   Linda is a teacher in an elementary school.
   Linda works in a bookstore and takes yoga classes.
   Linda is active in the feminist movement.
   Linda is a psychiatric social worker.
   Linda is a member of the League of Women Voters.
   Linda is a bank teller.
   Linda is an insurance salesperson.
   Linda is a bank teller and is active in the feminist movement.

After you've made your ranking, take a look at two pairs of statements in particular: *Bill plays jazz for a hobby* and *Bill is an accountant who plays jazz for a hobby,* and *Linda is a bank teller* and *Linda is a bank teller and is active in the feminist movement.* Which of the two statements have you ranked as more likely in each pair?

I am willing to bet that it was the second one in both cases. If it was, you'd be with the majority, and you would be making a big mistake.

This exercise was taken verbatim from a 1983 paper by Amos Tversky and Daniel Kahneman, to illustrate our present point: when it comes to separating crucial details from incidental ones, we often don't fare particularly well. When the researchers' subjects were presented with these lists, they repeatedly made the same judgment that I've just predicted you would make: that it was more likely that Bill was an accountant who plays jazz for a hobby than it was that he plays jazz for a hobby, and that it was more likely that Linda was a feminist bank teller than that she was a bank teller at all.

Logically, neither idea makes sense: a conjunction cannot be more likely than either of its parts. If you didn't think it likely that Bill played jazz or that Linda was a bank teller to begin with, you should not have altered that judgment just because you *did* think it probable that Bill was an accountant and Linda, a feminist. An unlikely element or event when combined with a likely one does not somehow magically become any more likely. And yet 87 percent and 85 percent of participants, for the Bill scenario and the Linda scenario, respectively, made that exact judgment, in the process committing the infamous conjunction fallacy.

They even made it when their choices were limited: if only the two relevant options (Linda is a bank teller or Linda is a feminist bank teller) were included, 85 percent of participants *still* ranked the conjunction as more likely than the single instance. Even when people were given the logic behind the statements, they sided with the incorrect resemblance logic (*Linda seems more like a feminist, so I will say it's more likely that she's a feminist bank teller*) over the correct extensional logic (feminist bank tellers are only a specific subset of bank tellers, so Linda must be a bank teller with a higher likelihood than she would be a feminist one in particular) in 65 percent of cases. We can all be presented with the same set of facts and features, but the conclusions we draw from them need not match accordingly.

Our brains weren't made to assess things in this light, and our failings here actually make a good amount of sense. When it comes to things like chance and probability, we tend to be naive reasoners (and as chance and probability play a large part in many of our deductions, it's no wonder that we often go astray). It's called probabilistic incoherence, and it all stems from that same pragmatic storytelling that we engage in so naturally and readily—a tendency that may go back to a deeper, neural explanation; to, in some sense, W.J. and the split brain.

Simply put, while probabilistic reasoning seems to be localized in the left hemisphere, deduction appears to activate mostly the right hemisphere. In other words, the neural loci for evaluating logical implications and those for looking at their empirical plausibility may be in opposite hemispheres—a cognitive architecture that isn't conducive to coordinating statement logic with the assessment of chance and probability. As a result, we aren't always good at integrating various demands, and we often fail to do so properly, all the while remaining perfectly convinced that we have succeeded admirably.

The description of Linda and feminist (and Bill and accountant) coincides so well that we find it hard to dismiss the match as anything but hard fact. What is crucial here is our understanding of how frequently something occurs in real life—and the logical, elementary notion that a whole simply can't be more likely than the sum of its parts. And yet we let the incidental descriptors color our minds so much that we overlook the crucial probabilities.

What we should be doing is something much more prosaic. We should be gauging how likely any separate occurrence actually is. In chapter three, I introduced the concept of base rates, or how frequently something appears in the population, and promised to revisit it when we discussed deduction. And that's because base rates, or our ignorance of them, are at the heart of deductive errors like the conjunction fallacy. They hamper observation, but where they really throw you off is in deduction, in moving from all of your observations to the conclusions they imply. Because here, selectivity—and selective ignorance—will throw you off completely.

To accurately cast Bill and Linda's likelihood of belonging to any of the professions, we need to understand the prevalence of accountants, bank tellers, amateur jazz musicians, active feminists, and the whole lot in the population at large. We can't take our protagonists out of context. We can't allow one potential match to throw off other information we might have.

So, how does one go about resisting this trap, sorting the details properly instead of being swept up in irrelevance?

Perhaps the pinnacle of Holmes's deductive prowess comes in a case that is less traditional than many of his London pursuits. Silver Blaze, the prize-winning horse of the story's title, goes missing days before the big Wessex Cup race, on which many a fortune ride. That same morning, his trainer is found dead some distance from the stable. His skull looks like it has been hit by some large, blunt object. The lackey who had been guarding the horse has been drugged and remembers precious little of the night's events.

The case is a sensational one: Silver Blaze is one of the most famous horses in England. And so, Scotland Yard sends Inspector Gregson to investigate. Gregson, however, is at a loss. He arrests the most likely suspect—a gentleman who had been seen around the stable the evening of the disappearance—but admits that all evidence is circumstantial and that the picture may change at any moment. And so, three days later, with no horse in sight, Sherlock Holmes and Dr. Watson make their way to Dartmoor.

Will the horse run the race? Will the trainer's murderer be brought to justice? Four more days pass. It is the morning of the race. Silver Blaze, Holmes assures the worried owner, Colonel Ross, will run. Not to fear. And run he does. He not only runs, but wins. And his trainer's murderer is identified soon thereafter.

We'll be returning to "Silver Blaze" several times for its insights into the science of deduction, but first let's consider how Holmes introduces the case to Watson.

"It is one of those cases," says Holmes, "where the art of the reasoner should be used rather for the sifting of details than for the acquiring of fresh evidence. The tragedy has been so uncommon, so complete, and of such personal importance to so many people that we are suffering from a plethora of surmise, conjecture, and hypothesis." In other words, there is too much information to begin with, too many details to be able to start making them into any sort of coherent whole, separating the crucial from the incidental. When so many facts are piled together, the task becomes increasingly problematic. You have a vast quantity of your own observations and data but also an even vaster quantity of potentially incorrect information from individuals who may not have observed as mindfully as you have.

Holmes puts the problem this way: "The difficulty is to detach the framework of fact—of absolute undeniable fact—from the embellishments of theorists and reporters. Then, having established ourselves upon this sound basis, it is our duty to see what inferences may be drawn and what are the special points upon which the whole mystery turns." In other words, in sorting through the morass of Bill and Linda, we would have done well to set clearly in our minds what were the actual facts, and what were the embellishments or stories of our minds.

When we pry the incidental and the crucial apart, we have to exercise the same care that we spent on observing to make sure that we have recorded accurately all of the impressions. If we're not careful, mindset, preconception, or subsequent turns can affect even what we think we observed in the first place.

In one of Elizabeth Loftus's classic studies of eyewitness testimony, participants viewed a film depicting an automobile accident. Loftus then asked each participant to estimate how fast the cars were going when the accident occurred—a classic deduction from available data. But here's the twist: each time she asked the question, she subtly altered the phrasing. Her description of the accident varied by verb: the cars *smashed, collided, bumped, contacted,* or *hit.* What Loftus found was that her phrasing had a drastic impact on subjects' memory. Not only did those who viewed the "smashed" condition estimate a higher speed than those who viewed the other conditions, but they were also far more likely to recall, one week later, having seen broken glass in the film, even though there was actually no broken glass at all.

It's called the misinformation effect. When we are exposed to misleading information, we are likely to recall it as true and to take it into consideration in our deductive process. (In the Loftus experiment, the subjects weren't even exposed to anything patently false, just misleading.) All the specific word choice does is act as a simple frame that impacts our line of reasoning and even our memory. Hence the difficulty, and the absolute necessity, that Holmes describes of learning to sift what is irrelevant (and all that is media conjecture) from the real, objective, hard facts—and to do so thinkingly and systematically. If you don't, you may find yourself remembering broken glass instead of the intact windshield you actually saw.

In fact, it's when we have more, not less, information that we should be most careful. Our confidence in our deductions tends to increase along with the number of details on which we base them—especially if one of those details makes sense. A longer list somehow seems more reasonable, even if we were to judge individual items on that list as less than probable given the information at hand. So when we see one element in a conjunction that seems to fit, we are likely to accept the full conjunction, even if it makes little sense to do so. Linda the feminist bank teller. Bill the jazz-playing accountant. It's perverse, in a way. The better we've observed and the more data we've collected, the more likely we are to be led astray by a single governing detail.

Similarly, the more incidental details we see, the less likely we are to home in on the crucial, and the more likely we are to give the incidental undue weight. If we are told a story, we are more likely to find it compelling and true if we are also given more details, even if those details are irrelevant to the story's truth. Psychologist Ruma Falk has noted that when a narrator adds specific, superfluous details to a story of coincidence (for instance, that two people win the lottery in the same small town), listeners are more likely to find the coincidence surprising and compelling.

Usually when we reason, our minds have a tendency to grab any information that seems to be related to the topic, in the process retrieving both relevant cues and those that seem somehow to be connected but may not actually matter. We may do this for several reasons: familiarity, or a sense that we've seen this before or should know something even when we can't quite put our finger on it; spreading activation, or the idea that the activation of one little memory node triggers others, and over time the triggered memories spread further away from the original; or simple accident or coincidence—we just happen to think of something while thinking about something else.

If, for example, Holmes were to magically emerge from the book and ask us, not Watson, to enumerate the particulars of the case at hand, we'd rummage through our memory (*What did I just read? Or was that the other case?*), take certain facts out of storage (*Okay: horse gone, trainer dead, lackey drugged, possible suspect apprehended. Am I missing anything?*), and in the process, likely bring up others that may not matter all that much (*I think I forgot to eat lunch because I was so caught up in the drama; it's like that time I was reading* The Hound of the Baskervilles *for the first time, and forgot to eat, and then my head hurt, and I was in bed, and . . .*).

If the tendency to over-activate and over-include isn't checked, the activation can spread far wider than is useful for the purpose at hand—and can even interfere with the proper perspective needed to focus on that purpose. In the case of Silver Blaze, Colonel Ross constantly urges Holmes to do more, look at more, consider more, to leave, in his words, "no stone unturned." Energy and activity, more is more; those are his governing principles. He is supremely frustrated when Holmes refuses, choosing instead to focus on the key elements that he has already identified. But Holmes realizes that to weed out the incidental, he should do anything *but* take in more and more theories and potentially relevant (or not) facts.

We need, in essence, to do just what the CRT teaches us: reflect, inhibit, and edit. Plug System Holmes in, check the tendency to gather detail thoughtlessly, and instead focus—thoughtfully—on the details we already have. All of those observations? We need to learn to divide them in our minds in order to maximize productive reasoning. We have to learn when *not* to think of them as well as when to bring them in. We have to learn to concentrate—reflect, inhibit, edit—otherwise we may end up getting exactly nowhere on any of the myriad ideas floating through our heads. Mindfulness and motivation are essential to success-ful deduction.

But essential never means simple, nor does it mean sufficient. Even with Silver Blaze, Holmes, as focused and motivated as he is, finds it dif-ficult to sift through all of the possible lines of thought. As he tells Wat-son once Silver Blaze is recovered, "I confess that any theories which I had formed from the newspaper reports were entirely erroneous. And yet there were indications there, had they not been overlaid by other details which concealed their true import." The separation of crucial and inci-dental, the backbone of any deduction, can be hard for even the best-trained minds. That's why Holmes doesn't run off based on his initial theories. He first does precisely what he urges us to do: lay the facts out in a neat row and proceed from there. Even in his mistakes, he is delibera-tive and Holmes-like, not letting System Watson act though it may well want to.

How does he do this? He goes at his own pace, ignoring everyone who urges haste. He doesn't let anyone affect him. He does what he needs to do. And beyond that he uses another simple trick. He tells Watson everything—something that occurs with great regularity throughout the Holmes canon (and you thought it was just a clever expository device!). As he tells the doctor before he delves into the pertinent observations, "nothing clears up a case so much as stating it to another person." It's the exact same principle we've seen in operation before: stating something through, out loud, forces pauses and reflection. It mandates mindfulness. It forces you to consider each premise on its logical merits and allows you to slow down your thinking so that you do not blunder into a feminist Linda. It ensures that you do not let something that is of real significance go by simply because it didn't catch your attention enough or fit with the causal story that you have (subconsciously, no doubt) already created in your head. It allows your inner Holmes to listen and forces your Watson to pause. It allows you to confirm that you've actually understood, not just thought you understood because it seemed right.

Indeed, it is precisely in stating the facts to Watson that Holmes real-izes the thing that will allow him to solve the case. "It was while I was in the carriage, just as we reached the trainer's house, that the immense significance of the curried mutton occurred to me." The choice of a dinner is easy to mistake for triviality, until you state it along with everything else and realize that the dish was perfectly engineered to hide the smell and taste of powdered opium, the poison that was used on the stable boy. Someone who didn't know the curried mutton was to be served would never risk using a poison that could be tasted. The culprit, then, is someone who knew what was for dinner. And that realization prompts Holmes to his famous conclusion: "Before deciding that question I had grasped the significance of the silence of the dog, for one true inference invariably suggests others." Start on the right track, and you are far more likely to remain there.

While you're at it, make sure you are recalling *all* of your observations, all of the possible permutations that you've thought up in your imaginative space, and avoiding those instances that are not part of the picture. You can't just focus on the details that come to mind most easily or the ones that seem to be representative or the ones that seem to be most salient or the ones that make the most intuitive sense. You have to dig deeper. You would likely never judge Linda a likely bank teller from her description, though you very well might judge her a likely feminist. Don't let that latter judgment color what follows; instead, proceed with the same logic that you did before, evaluating each element separately and objectively as part of a consistent whole. A likely bank teller? Absolutely not. And so, a feminist one? Even less probable.

You have to remember, like Holmes, all of the details about Silver Blaze's disappearance, stripped of all of the papers' conjectures and the theories your mind may have inadvertently formed as a result. Never would Holmes call Linda a feminist bank teller, unless he was first certain that she was a bank teller.

The Improbable Is Not Impossible

In *The Sign of Four*, a robbery and murder are committed in a small room, locked from the inside, on the top floor of a rather large estate. How in the world did the criminal get inside to do the deed? Holmes enumerates the possibilities: "The door has not been opened since last night," he tells Watson. "Window is snibbed on the inner side. Framework is solid. No hinges at the side. Let us open it. No water-pipe near. Roof quite out of reach."

Then how to possibly get inside? Watson ventures a guess: "The door is locked, the window is inaccessible. Was it through the chimney?"

No, Holmes tells him. "The grate is much too small. I had already considered that possibility."

"How then?" asks an exasperated Watson.

"You will not apply my precept." Holmes shakes his head. "How often have I said to you that when you have eliminated the impossible, whatever remains, *however improbable*, must be the truth? We know that he did not come through the door, the window, or the chimney. We also know that he could not have been concealed in the room, as there is no concealment possible. Whence, then, did he come?"

And then, at last, Watson sees the answer: "He came from the hole in the roof." And Holmes's reply, "Of course he did. He must have done so," makes it seem the most logical entrance possible.

It isn't, of course. It is highly improbable, a proposition that most people would never consider, just as Watson, trained as he is in Holmes's approach, failed to do without prompting. Just like we find it difficult to separate the incidental from the truly crucial, so, too, we often fail to consider the improbable—because our minds dismiss it as impossible before we even give it its due. And it's up to System Holmes to shock us out of that easy narrative and force us to consider that something as unlikely as a rooftop entrance may be the very thing we need to solve our case.

Lucretius called a fool someone who believes that the tallest mountain that exists in the world and the tallest mountain he has ever observed are one and the same. We'd probably brand someone who thought that way foolish as well. And yet we do the same thing every single day. Author and mathematician Nassim Taleb even has a name for it, inspired by the Latin poet: the Lucretius underestimation. (And back in Lucretius's day, was it so strange to think that your world was limited to what you knew? In some ways, it's smarter than the mistakes we make today given the ease of knowledge at our disposal.)

Simply put, we let our own personal past experience guide what we perceive to be possible. Our repertoire becomes an anchor of sorts; it is our reasoning starting point, our place of departure for any further thoughts. And even if we try to adjust from our egocentric perspective, we tend not to adjust nearly enough to matter, remaining stubbornly skewed in a self-directed approach. It's our storytelling proclivity in another guise: we imagine stories based on the ones we've experienced, not the ones we haven't.

Learning of historical precedent as well matters little, since we don't learn in the same way from description as we do from experience. It's something known as the description-experience gap. Perhaps Watson had read at one time or another about a daring rooftop entrance, but because he has never had direct experience from it, he will not have processed the information in the same way and is not likely to use it in the same manner when trying to solve a problem. Lucretius's fool? Having read of high peaks, he may *still* not believe they exist. *I want to see them with my own two eyes,* he'll say. *What am I, some kind of fool?* Absent a direct precedent, the improbable seems so near impossible that Holmes's maxim falls by the wayside.

And yet distinguishing the two is an essential ability to have. For, even if we have successfully separated the crucial from the incidental, even if we've gathered all of the facts (and their implications) and have focused on the ones that are truly relevant, we are lost if we don't let our minds think of the roof, however unlikely it is, as a possible entry point into a room. If, like Watson, we dismiss it out of hand—or fail to even think about—we will never be able to deduce those alternatives that would flow directly from our reasoning if only we'd let them.

We use the best metric of the future—the past. It's natural to do so, but that doesn't mean it's accurate. The past doesn't often make room for the improbable. It constrains our deduction to the known, the likely, the probable. And who is to say that the evidence, if taken together and properly considered, doesn't lead to an alternative beyond these realms?

Let's go back for a moment to "Silver Blaze." Sherlock Holmes emerges triumphant, it's true—the horse is found, as is the trainer's murderer— but not after a delay that is uncharacteristic of the great detective. He is late to the investigation (three days late, to be specific), losing valuable

time at the scene. Why? He does just what he reprimands Watson for doing: he fails to apply the precept that the improbable is not yet the impossible, that it must be considered along with the more likely alternatives.

As Holmes and Watson head to Dartmoor to help with the investigation, Holmes mentions that on Tuesday evening both the horse's owner
and Inspector Gregson had telegraphed for his assistance on the case.
The flummoxed Watson responds, "Tuesday evening! And this is Thursday morning. Why didn't you go down yesterday?" To which Holmes answers, "Because I made a blunder, my dear Watson—which is, I am
afraid, a more common occurrence than anyone would think who only
knew me through your memoirs. The fact is that I could not believe it
possible that the most remarkable horse in England could long remain
concealed, especially in so sparsely inhabited a place as the north of
Dartmoor."

Holmes has dismissed the merely improbable as impossible and has
failed to act in a timely fashion as a result. In so doing, he has reversed
the usual Holmes-Watson exchange, making Watson's reprimand un-
characteristically well warranted and on point.

Even the best and sharpest mind is necessarily subject to its owner's
unique experience and world perception. While a mind such as Holmes's
is, as a rule, able to consider even the most remote of possibilities, there
are times when it, too, becomes limited by preconceived notions, by what
is available to its repertoire at any given point. In short, even Holmes is
limited by the architecture of his brain attic.

Holmes sees a horse of exceptional appearance missing in a rural
area. Everything in his experience tells him it can't go missing for long.
His logic is as follows: if the horse is the most remarkable such animal in
the whole of England, then how could it go under the radar in a remote
area where hiding places are limited? Surely someone would notice the
beast, dead or alive, and make a report. And that would be perfect deduction from the facts, if it happened to be true. But it is Thursday, the horse
has been missing since Tuesday, and the report has failed to come. What
is it then that Holmes failed to take into account?

A horse couldn't remain concealed *if it could still be recognized as that
horse*. The possibility of disguising the animal doesn't cross the great de-

tective's mind; if it had, surely he wouldn't have discounted the likeli-hood of the animal remaining hidden. What Holmes sees isn't just what there is; he is also seeing what he knows. Were we to witness something that in no way fit with past schemas, had no counterpart in our memory, we would likely not know how to interpret it—or we may even fail to see it altogether, and instead see what we were expecting all along.

Think of it as a complex version of any one of the famous Gestalt demonstrations of visual perception, whereby we are easily able to see one thing in multiple ways, depending on the context of presentation.

For instance, consider this picture:

Do you see the middle figure as a *B* or a 13? The stimulus remains the same, but what we see is all a matter of expectation and context. A dis-guised animal? Not in Holmes's repertoire, however vast it might be, and so he does not even consider the possibility. Availability—from experi-ence, from contextual frames, from ready anchors—affects deduction. We wouldn't deduce a *B* if we took away the *A* and *C*, just like we'd never deduce a 13 were the 12 and 14 to be removed. It wouldn't even cross our minds, even though it is highly possible, merely improbable given the context. But if the context were to shift slightly? Or if the missing row were to be present, only hidden from our view? That would change the picture, but it wouldn't necessarily change the choices we consider.

This raises another interesting point: not only does our experience affect what we consider possible, but so, too, do our expectations. Holmes was *expecting* Silver Blaze to be found, and as a result he viewed his evi-dence in a different light, allowing certain possibilities to go unexamined.

Demand characteristics rear their ugly head yet again; only this time they take the guise of the confirmation bias, one of the most prevalent mistakes made by novice and experienced minds alike.

From early childhood, we seem to be susceptible to forming confirmatory biases, to deciding long before we actually decide and dismissing the improbable out of hand as impossible. In one early study of the phenomenon, children as young as third grade were asked to identify which features of sports balls were important to the quality of a person's serve. Once they made up their minds (for instance, size matters but color does not), they either altogether failed to acknowledge evidence that was contrary to their preferred theory (such as the actual importance of color, or the lack thereof of size) or considered it in a highly selective and distorted fashion that explained away anything that didn't correspond to their initial thought. Furthermore, they failed to generate alternative theories unless prompted to do so, and when they later recalled both the theory and the evidence, they misremembered the process so that the evidence became much more consistent with the theory than it had been in reality. In other words, they recast the past to better suit their own view of the world.

As we age, it only gets worse—or at the very least it doesn't get any better. Adults are more likely to judge one-sided arguments as superior to those that present both sides of a case, and more likely to think that such arguments represent good thinking. We are also more likely to search for confirming, positive evidence for hypotheses and established beliefs even when we are not actually invested in those hypotheses. In a seminal study, researchers found that participants tested a concept by looking only at examples that would hold if that concept were correct—and failed to find things that would show it to be incorrect. Finally, we exhibit a remarkable asymmetry in how we weigh evidence of a hypothesis: we tend to overweight any positive confirming evidence and underweight any negative disconfirming evidence—a tendency that professional mind readers have exploited for ages. We see what we are looking for.

In these final stages of deduction, System Watson will still not let us go. Even if we *do* have all the evidence, as we surely will by this point in the process, we might *still* theorize before the evidence, in letting our experience and our notion of what is and is not possible color how we see and apply that evidence. It's Holmes disregarding the signs in "Silver Blaze" that would point him in the right direction because he doesn't consider it possible that the horse could remain undetected. It's Watson disregarding the roof as an option for entrance because he doesn't consider it possible that someone could enter a room in that fashion. We might have all the evidence, but that doesn't mean when we reason, we'll take into account that all of the evidence is objective, intact, and in front of us.

But Holmes, as we know, does manage to catch and correct his error—or have it caught for him, with the failure of the horse to materialize. And as soon as he allows that improbable possibility to become possible, his entire evaluation of the case and the evidence changes and falls into place. And off he and Watson go to find the horse and save the day. Likewise, Watson is able to correct his incomprehension when prompted to do so. Once Holmes reminds him that however improbable something may be, it must still be considered, he right away comes up with the alternative that fits the evidence—an alternative that just a moment ago he had dismissed entirely.

The improbable is not yet impossible. As we deduce, we are too prone to that satisficing tendency, stopping when something is good enough. Until we have exhausted the possibilities and are sure that we have done so, we aren't home clear. We must learn to stretch our experience, to go beyond our initial instinct. We must learn to look for evidence that both confirms and disconfirms and, most important, we must try to look beyond the perspective that is the all too natural one to take: our own.

We must, in short, go back to that CRT and its steps; reflect on what our minds *want* to do; inhibit what doesn't make sense (here, asking whether something is truly impossible or merely unlikely); and edit our approach accordingly. We won't always have a Holmes prompting us to do so, but that doesn't mean we can't prompt ourselves, through that very mindfulness that we've been cultivating. While we may still be tempted to act first and think later, to dismiss options before we've even considered them, we can at least recognize the general concept: think first, act later, and try our utmost to approach every decision with a fresh mind.

The necessary elements are all there (at least if you've done your observational and imaginative work). The trick is in what you do with them. Are you using all available evidence, and not just what you happen to remember or think of or encounter? Are you giving it all the same weight, so that you are truly able to sift the crucial from the incidental instead of being swayed by some other, altogether irrelevant factors? Are you laying each piece out in a logical sequence, where each step implies the next and each factor is taken to its conclusion, so that you don't fall victim to the mistake of thinking you've thought it through when you've done no such thing? Are you considering all logical paths—even those that may seem to you to be impossible? And finally: are you focused and motivated? Do you remember what the problem was that got you there in the first place—or have you been tempted off course, or off to some other problem, without really knowing how or why?

I first read Sherlock Holmes in Russian because that was the language of my childhood and of all of my childhood books. Think back to the clues I've left for you. I've told you that my family is Russian, and that both my sister and I were born in the Soviet Union. I've told you that the stories were read to me by my dad. I've told you that the book in question was old—so old that I wondered if his dad had, in turn, read it to him. In what other language could it have possibly been, once you see everything laid out together? But did you stop to consider that as you were seeing each piece of information separately? Or did it not even cross your mind because of its . . . improbability? Because Holmes is just so, well, English?

It doesn't matter that Conan Doyle wrote in English and that Holmes himself is so deeply ingrained in the consciousness of the English language. It doesn't matter that I now read and write in English just as well as I ever did in Russian. It doesn't matter that you may have never encountered a Russian Sherlock Holmes or even considered the likelihood of his existence. All that matters is what the premises are and where they take you if you let them unwind to their logical conclusion, whether or not that is the place that your mind had been gearing to go.

SHERLOCK HOLMES FURTHER READING

"'*Elementary,*' *said he.*" "*I smoked several pipes over them, trying to sepa-rate those which were crucial from others which were merely incidental.*" from *The Memoirs of Sherlock Holmes,* "The Crooked Man," p. 138.

"*Every instinct that I possess cries out against it.*" from *The Return of Sherlock Holmes,* "The Adventure of the Abbey Grange," p. 1158.

"*It is one of those cases where the art of the reasoner should be used rather for the sifting of details . . .*" "*I confess that any theories which I had formed from the newspaper reports were entirely erroneous.*" from *The Memoirs of Sherlock Holmes,* "Silver Blaze," p. 1.

"*How often have I said to you that when you have eliminated the im-possible, whatever remains,* however improbable, *must be the truth?*" from *The Sign of Four,* chapter 6: Sherlock Holmes Gives a Demonstra-tion, p. 41.

CHAPTER SIX

Maintaining the Brain Attic: Education Never Stops

A lodger's behavior has been markedly unusual. His landlady, Mrs. Warren, hasn't seen him a single time over a period of ten days. He remains always in his room—save for the first evening of his stay, when he went out and returned late at night—pacing back and forth, day in, day out. What's more, when he needs something, he prints a single word on a scrap of paper and leaves it outside: SOAP. MATCH. DAILY GAZETTE. Mrs. Warren is alarmed. She feels that something must be wrong. And so she sets off to consult Sherlock Holmes.

At first, Holmes has little interest in the case. A mysterious lodger hardly seems worth investigating. But little by little, the details begin to grow intriguing. First, there is the business of the printed words. Why not write them normally instead? Why choose such a cumbersome, unnatural all-caps means of communication? Then there is the cigarette, which Mrs. Warren has helpfully brought along: while the landlady has assured Holmes that the mystery man has a beard and mustache, Holmes asserts that only a clean-shaven man could have smoked the cigarette in question. Still, it is not much to go on, so the detective tells Mrs. Warren to report back "if anything fresh occurs."

And something does occur. The following morning, Mrs. Warren returns to Baker Street with the following exclamation: "It's a police matter, Mr. Holmes! I'll have no more of it!" Mr. Warren, the landlady's husband, has been attacked by two men, who put a coat over his head and threw him into a cab, only to release him, roughly an hour later. Mrs. Warren blames the lodger and resolves to have him out that very day.

Not so fast, says Holmes. "Do nothing rash. I begin to think that this affair may be very much more important than appeared at first sight. It is clear now that some danger is threatening your lodger. It is equally clear that his enemies, lying in wait for him near your door, mistook your husband for him in the foggy morning light. On discovering their mistake they released him."

That afternoon, Holmes and Watson travel to Great Orme Street, to glimpse the identity of the guest whose presence has caused such a stir. Soon enough, they see her—for it is, in fact, a she. Holmes's conjecture had been correct: a substitution of lodger has been made. "A couple seek refuge in London from a very terrible and instant danger. The measure of that danger is the rigour of their precautions," Holmes explains to Watson.

> "The man, who has some work which he must do, desires to leave the woman in absolute safety while he does it. It is not an easy problem, but he solved it in an original fashion, and so effectively that her presence was not even known to the landlady who supplies her with food. The printed messages, as is now evident, were to prevent her sex being discovered by her writing. The man cannot come near the woman, or he will guide their enemies to her. Since he cannot communicate with her direct, he has recourse to the agony column of a paper. So far all is clear."

But to what end? Watson wants to know. Why the secrecy and the danger? Holmes presumes that the matter is one of life and death. The attack on Mr. Warren, the lodger's look of horror when she suspects someone might be looking at her, everything points to a sinister cast.

Why, then, asks Watson, should Holmes continue to investigate? He has solved Mrs. Warren's case—and the landlady herself would like nothing more than to force the lodger out of the boardinghouse. Why involve himself further, especially if the case is as risky as it sounds? It would be easy enough to leave and let events take their course. "What have you to gain from it?" he asks the detective.

Holmes has a ready answer:

> "What, indeed? It is art for art's sake. Watson, I suppose when you doctored you found yourself studying cases without a thought of a fee?"

> "For my education, Holmes."

"Education never ends, Watson. It is a series of lessons with the greatest for the last. This is an instructive case. There is neither money nor credit in it, and yet one would wish to tidy it up. When dusk comes we should find ourselves one stage advanced in our investigation."

It doesn't matter to Holmes that the initial goal has been attained. It doesn't matter that the further pursuit of the matter is dangerous in the extreme. You don't just abandon something when your original goal is complete, if that something has proven itself more complex than it may have seemed at first. The case is instructive. If nothing else, there is still more to learn. When Holmes says that education never ends, his message to us isn't as one-dimensional as it may seem. Of course it's good to keep learning: it keeps our minds sharp and alert and prevents us from settling in our ways. But for Holmes, education means something more. Education in the Holmesian sense is a way to keep challenging yourself and questioning your habits, of never allowing System Watson to take over altogether—even though he may have learned a great deal from System Holmes along the way. It's a way of constantly shaking up our habitual behaviors, and of never forgetting that, no matter how expert we think we are at something, we must remain mindful and motivated in everything we do.

This whole book has stressed the necessity of practice. Holmes got to where he is because of constantly practicing those mindful habits of thought that form the core of his approach to the world. As we practice, however, as things become more and more simple and second nature, they move into the purview of System Watson. Even though the habits may now be Holmesian ones, they have all the same become habits, things we do as a matter of course—and therefore, if we're not careful, mindlessly. It's when we take our thinking for granted and stop paying attention to what is actually going on in our brain attic that we are prone to mess up, even if that attic is now the most streamlined and polished place you ever saw. Holmes must keep challenging himself lest he succumb to the very same thing. For even though his mindful habits are sharp indeed, even they can lead him astray if he doesn't keep applying them. If we don't keep challenging our habits of thought, we risk letting the mindfulness we've so carefully cultivated slip back into its pre-Holmesian, mindless existence.

It's a difficult task, and our brain, as usual, is of little help. When we feel like we've completed something worthwhile, be it a simple task like cleaning up a pesky closet, or something a bit more involved, like the resolution of a mystery, our Watson brain would like nothing better than to rest, to reward itself for a job well done. Why go further if you've done what you've set out to do?

Human learning is largely driven by something known as the reward prediction error (RPE). When something is more rewarding than expected—*I made the left turn! I didn't hit the cone!* in the case of learning to drive—the RPE leads to a release of dopamine into the brain. That release occurs frequently when we begin to learn something new. With each step, it is easy to see gratifying results: we begin to understand what we're doing, our performance improves, we make fewer mistakes. And each point of accomplishment *does* actually entail some gain for us. Not only are we performing better (which presumably will make us happy) but our brain is being rewarded for its learning and improvement.

But then, all of a sudden, it stops. It's no longer surprising that I can drive smoothly. It's no longer surprising that I'm not making mistakes on my typing. It's no longer surprising that I can tell that Watson came from Afghanistan. I know I'll be able to do it before I actually do it. And so there's no RPE. No RPE, no dopamine. No pleasure. No need for further learning. We've achieved a suitable plateau and we decide—on a neural level as well as a conscious one—that we've learned all we need to know.

The trick is to train your brain to move *past* that point of immediate reward, to find the uncertainty of the future rewarding in itself. It's not easy—for as I've said before, future uncertainty is precisely the thing we *don't* much like. Far better to reap the benefits now, and bask in the dopamine ride and its aftereffects.

Inertia is a powerful force. We are creatures of habit—and not just observable habits, such as, for instance, always putting on the TV when we walk into our living room after work, or opening the fridge just to see what's in there, but thought habits, predictable loops of thinking that, when triggered, go down a predictable path. And thought habits are tough to break.

One of the most powerful forces of choice is the default effect—the tendency, as we've already discussed, to choose the path of least resistance, going with what is in front of us as long as that is a reasonable enough option. We see it playing out all the time. At work, employees tend to contribute to retirement plans when the contribution is the default and to stop contributing—even when matched generously by employers—when they need to opt in. Countries where organ donation is the default (each person is an organ donor unless he actively specifies that he doesn't want to be) have significantly higher percentages of donors than countries where donors must opt in. Effectively, when given a choice between doing something and nothing, we choose the nothing—and tend to forget that that, too, is doing something. But it's doing something quite passive and complacent, the polar opposite of the active engagement that Holmes always stresses.

And here's the odd thing: the better we are, the better we've become, the more we've learned, the more powerful is the urge to just rest already. We feel like we somehow deserve it, instead of realizing that it is the greatest disservice we could possibly do ourselves.

We see this pattern playing itself out not merely at the individual levels but throughout organizations and corporations. Think about how many companies have produced breakthrough innovations only to find themselves swamped by competitors and left behind a few years later. (Consider, for instance, Kodak or Atari or RIM, creator of the Black-Berry.) And this tendency isn't limited to the business world. The pattern of spectacular innovation followed by just as spectacular stagnation describes a more general trend that occurs in academia, the military, and almost any industry or profession you can name. And it's all rooted in how our brain's reward system is set up.

Why are these patterns so common? It goes back to those default effects, that inertia, on a much broader level: to the entrenchment of habit. And the more rewarded a habit is, the harder it is to break. If a gold star on a spelling test is enough to send dopamine firing in a child's brain, just imagine what multibillion-dollar success, soaring market shares, bestseller or award-winning or tenure-worthy academic fame can do.

·   ·   ·

We've spoken before about the difference between short- and long-term memory, those things we hold on to just briefly before letting them go and those we store in our brain attic more permanently. The latter seems to come in two flavors (though its exact mechanisms are still being investigated): *declarative*, or explicit memory, and *procedural*, or implicit memory. Think of the first as a kind of encyclopedia of knowledge about events (episodic memory) or facts (semantic memory) or other things that you can recall explicitly. Each time you learn a new one, you can write it down under its own, separate entry. Then, if you're asked about that particular entry, you can flip to that page of the book and—if everything goes well and you've written it down properly and the ink hasn't faded—retrieve it. But what if something can't be written down per se? What if it's just something you kind of feel or know how to do? Then you've moved to the realm of procedural, or implicit memory. Experience. It's no longer as easy as an encyclopedia entry. If I were to ask you about it directly, you may not be able to tell me, and it might even disrupt the very thing I was asking you about. The two systems are not entirely separate and do interact quite a bit, but for our purposes you can think of them as two different types of information that are stored in your attic. Both are there, but they are not equally conscious or accessible. And you can move from one to the other without quite realizing you've done so.

Think of it like learning to drive a car. At first, you explicitly remember everything you need to do: turn the key, check your mirrors, take the car out of park, and on and on. You have to consciously execute each step. But soon you stop thinking of the steps. They become second nature. And if I were to ask you what you were doing, you might not even be able to tell me. You've moved from explicit to implicit memory, from active knowledge to habit. And in the realm of implicit memory, it is far more difficult to improve consciously or to be mindful and present. You have to work much harder to maintain the same level of alertness as when you were just learning. (That's why so much learning reaches what K. Anders Ericsson terms a plateau, a point beyond which we can't seem to improve. As we'll find out, that is not actually true, but it is difficult to overcome.)

When we are first learning, we are in the realm of declarative, or explicit memory. That's the memory that is encoded in the hippocampus and then consolidated and stored (if all goes well) for future use. It's the memory we use as we memorize dates in history or learn the steps of a new procedure at work. It's also the memory I tried to use in memorizing the numbers of stairs in all possible houses (and failed at miserably) when I completely misunderstood Holmes's point, and the memory we use as we try to embrace Holmes's thought process step-by-step, so that we can begin to approximate his powers of insight.

But it's not the same memory that Holmes uses when he does the same thing. He has already mastered those steps of thought. To him they have become second nature. Holmes doesn't need to think about thinking, in the proper fashion; he does it automatically—just as we automatically default to our inner Watson because it's what we've *learned* to do and are now unlearning.

Until we unlearn, what to Holmes is effortless couldn't be more effortful to our Watson selves. We must stop Watson at every point and ask instead the opinion of Holmes. But as we practice this more and more, as we force ourselves to observe, to imagine, to deduce over and over and over—and to do it even in those circumstances where it may seem silly, like deciding what to have for lunch—a change takes place. Suddenly, things flow a little more smoothly. We proceed a little more quickly. It feels a little more natural, a little more effortless.

In essence, what is happening is that we are switching memory systems. We are moving from the explicit to the implicit, the habitual, the procedural. Our thinking is becoming akin to the memory that we have when we drive, when we ride a bike, when we complete a task that we've done countless times. We've gone from being goal directed (in the case of thinking, of consciously going through Holmes's steps, making sure to execute each one properly) to being automated (we no longer have to think about the steps; our minds go through them as a matter of course). From something that is based largely on effortful memory to something that triggers that dopamine reward system without our necessarily realizing it (think of an addict's behavior—an extreme example). And here allow me to repeat myself, because it bears repeating: the more rewarded something is, the quicker it will become a habit, and the harder it will be to break.

Bringing Habits Back from Mindlessness into Mindfulness

"The Adventure of the Creeping Man" takes place when Holmes and Watson no longer live together. One September evening, Watson receives a message from his former flatmate. "Come at once if convenient," it reads. "If inconvenient come all the same." Clearly, Holmes wants to see the good doctor—and as promptly as possible. But why? What could Watson have that Holmes so urgently needs, that can't wait or be communicated by message or messenger? If you think back on their time together, it's not clear that Watson has ever served a role much beyond that of faithful supporter and chronicler. Surely, he was never the one to solve the crime, come upon the key insight, or influence the case in any meaningful way. Surely, Sherlock Holmes's summons now couldn't be all that urgent—a message that is meant to ask for Watson's aid in solving a case.

But that is precisely what it is. As it turns out, Watson is—and has long been—far, far more than chronicler and friend, faithful companion and moral supporter. Watson is, in fact, part of the reason that Sherlock Holmes has managed to remain as sharp and ever mindful as he has been for as long as he has. Watson *has* been essential (indeed, irreplaceable) in solving a case, and will continue to be so, again and again. And soon, you will see precisely why that is.

Habit is useful. I'll even go a step further and say that habit is essential. It frees us up cognitively to think of broader, more strategic issues instead of worrying about the nitty-gritty. It allows us to think on a higher level and an altogether different plain than we would otherwise be able to do. In expertise lies great freedom and possibility.

On the other hand, habit is also perilously close to mindlessness. It is very easy to stop thinking once something becomes easy and automatic. Our effortful journey to attain the Holmesian habits of thought is goal directed. We are focused on reaching a future reward that comes of learning to think mindfully, of making better, more informed, and more thorough choices, of being in control of our minds instead of letting them control us. Habits are the opposite. When something is a habit, it has moved from the mindful, motivated System Holmes brain to the mindless, unthinking System Watson brain, which possesses all of those biases and heuristics, those hidden forces that begin to affect your behavior without your knowledge. You've stopped being aware of it, and because of that, you are far less able to pay attention to it.

And yet what about Sherlock Holmes? How does he manage to stay mindful? Doesn't that mean that habits need not be incompatible with mindfulness?

Let's go back to Holmes's urgent message to Watson, his call to come no matter how inconvenient the visit might be. Watson knows exactly why he is being called upon—though he may not realize just how essential he is. Holmes, says Watson, is "a man of habits, narrow and concentrated habits, and I had become one of them. As an institution I was like the violin, the shag tobacco, the old black pipe, the index books." And what, precisely, is the role of Watson-as-an-institution? "I was a whetstone for his mind. I stimulated him. He liked to think aloud in my presence. His remarks could hardly be said to be made to me—many of them would have been as appropriately addressed to his bedstead—but none the less, having formed the habit, it had become in some way helpful that I should register and interject." And that's not all. "If I irritated him by a certain methodical slowness in my mentality," Watson continues, "that irritation served only to make his own flame-like intuitions and impressions flash up the more vividly and swiftly. Such was my humble rôle in our alliance."

Holmes has other ways, to be sure—and Watson's role is but a component of a wider theme, as we'll soon see—but Watson is an irreplaceable tool in Holmes's multidimensional arsenal, and his function as tool (or institution, if you'd prefer) is to make sure that Holmes's habits of thought do not fall into mindless routine, that they remain ever mindful, ever present, and ever sharpened.

Earlier we talked about learning to drive and the danger we face when we've become proficient enough that we stop thinking about our actions, and so may find our attention drifting, our minds shifting into mindlessness. If all is as usual, we'd be fine. But what if something went awry? Our reaction time wouldn't be nearly as quick as it had been in the initial learning stages when we had focused on the road.

But what if we were forced to really think about our driving once more?

Someone taught us how to drive, and we might be called upon to teach someone else. If we are, we would be wise indeed to take up the challenge. When we talk something through to another person, break it down for his understanding, not only are we once again forced to pay attention to what we're doing, but we might even see our own driving improving. We might see ourselves thinking of the steps differently and becoming more mindful of what we're doing as we do it—if only to set a good example. We might see ourselves looking at the road in a fresh way, to be able to formulate what it is that our novice driver needs to know and notice, how he should watch and react. We might see patterns emerge that we hadn't taken into account—or been able to see, really—the first time around, when we were so busy mastering the composite steps. Not only will our cognitive resources be freer to see these things, but we will be present enough to take advantage of the freedom.

Likewise, Holmes. It's not just in "The Adventure of the Creeping Man" that he needs Watson's presence. Notice how in each case he is always teaching his companion, always telling him how he reached this or that conclusion, what his mind did and what path it took. And to do that, he must reflect back on the thought process. He must focus back in on what has become habit. He must be mindful of even those conclusions that he reached mindlessly, like knowing why Watson came from Afghanistan. (Though, as we've already discussed, Holmesian mindlessness is far different from Watsonian.) Watson prevents Holmes's mind from forgetting to think about those elements that come naturally.

What's more, Watson serves as a constant reminder of what errors are possible. As Holmes puts it, "In noting your fallacies I was occasionally guided towards the truth." And that is no small thing. Even in asking the smallest questions, ones that seem entirely obvious to Holmes, Watson nevertheless forces Holmes to look twice at the very obviousness of the thing, to either question it or explain why it is as plain as all that. Watson is, in other words, indispensable.

And Holmes knows it well. Look at his list of external habits: the violin, the tobacco and pipe, the index book. Each of his habits has been chosen mindfully. Each facilitates thought. What did he do pre-Watson? Whatever it was, he certainly realized very quickly that a post-Watson world was far preferable. "It may be that you are not yourself luminous," he tells Watson, not altogether unkindly on one occasion, "but you are a conductor of light. Some people without possessing genius have a re-markable power of stimulating it. I confess, my dear fellow, that I am very much in your debt." In his debt he most certainly is.

The greats don't become complacent. And that, in a nutshell, is Holmes's secret. Even though he doesn't need anyone to walk him through the sci-entific method of the mind—he may as well have invented the thing—he nevertheless keeps challenging himself to learn more, to do things better, to improve, to tackle a case or an angle or an approach that he has never seen in the past. Part of this goes back to his constant enlistment of Wat-son, who challenges him, stimulates him, and forces him to never take his prowess for granted. And another part goes to the choice of the cases themselves. Remember, Holmes doesn't take on just any case. He takes on only those that interest him. It's a tricky moral code. He doesn't take his cases merely to reduce crime but to challenge some aspect of his thinking. The commonplace criminal need not apply.

But either way, whether in cultivating Watson's companionship or in choosing the harder, more exceptional case over the easier one, the mes-sage is the same: keep feeding the need to learn and to improve. At the end of "The Red Circle" Holmes finds himself face-to-face with Inspector Gregson, who turns out to have been investigating the very case that Holmes decides to pursue after his initial work is done. Gregson is per-plexed to the extreme. "But what I can't make head or tail of, Mr. Holmes, is how on earth *you* got yourself mixed up in the matter," he says.

Holmes's response is simple. "Education, Gregson, education. Still seeking knowledge at the old university." The complexity and unrelated-ness of this second crime do the opposite of deterring him. They engage him and invite him to learn more.

In a way, that, too, is a habit, of never saying no to more knowledge, as scary or as complicated as it may be. The case in question is "a specimen of the tragic and grotesque," as Holmes says to Watson. And as such, it is well worthy of pursuit.

We, too, must resist the urge to pass on a difficult case, or to give in to the comfort of knowing we've already solved a crime, already accomplished a difficult task. Instead, we have to embrace the challenging, even when it is far easier not to. Only by doing so can we continue throughout our lives to reap the benefits of Holmesian thinking.

The Perils of Overconfidence

But how do we make sure we don't fall victim to overly confident thinking, thinking that forgets to challenge itself on a regular basis? No method is foolproof. In fact, thinking it foolproof is the very thing that might trip us up. Because our habits have become invisible to us, because we are no longer learning actively and it doesn't seem nearly as hard to think well as it once did, we tend to forget how difficult the process once was. We take for granted the very thing we should value. We think we've got it all under control, that our habits are still mindful, our brains still active, our minds still constantly learning and challenged—especially since we've worked so hard to get there—but we have instead replaced one, albeit far better, set of habits with another. In doing so we run the risk of falling prey to those two great slayers of success: complacency and overconfidence. These are powerful enemies indeed. Even to someone like Sherlock Holmes.

Consider for a moment "The Yellow Face," one of the rare cases where Holmes's theories turn out to be completely wrong. In the story, a man named Grant Munro approaches Holmes to uncover the cause of his wife's bizarre behavior. A cottage on the Munros' property has recently acquired new tenants, and strange ones at that. Mr. Munro glimpses one of its occupants and remarks that "there was something unnatural and inhuman about the face." The very sight of it chills him.

But even more surprising than the mystery tenants is his wife's response to their arrival. She leaves the house in the middle of the night, lying about her departure, and then visits the cottage the next day, extracting a promise from her husband that he will not try to pursue her inside. When she goes a third time, Munro follows, only to find the place deserted. But in the same room where he earlier saw the chilling face, he finds a photograph of his wife.

What ever is going on? "There's blackmail in it, or I am much mistaken," proclaims Holmes. And the blackmailer? "The creature who lives in the only comfortable room in the place and has her photograph above his fireplace. Upon my word, Watson, there is something very attractive about that livid face at the window, and I would not have missed the case for worlds."

Watson is intrigued at these tidbits. "You have a theory, then?" he asks.

"Yes, a provisional one," Holmes is quick to reply. "But," he adds, "I shall be surprised if it does not turn out to be correct. This woman's first husband is in that cottage."

But this provisional theory proves incorrect. The occupant of the cottage is not Mrs. Munro's first husband at all, but her daughter, a daughter of whose existence neither Mr. Munro nor Holmes had any prior knowledge. What had appeared to be blackmail is instead simply the money that enabled the daughter and the nanny to make the passage from America to England. And the face that had seemed so unnatural and inhuman was that way because it was, indeed, just that. It was a mask, designed to hide the little girl's black skin. In short, Holmes's wonderings have ended up far from the truth. How could the great detective have gone so wrong?

Confidence in ourselves and in our skills allows us to push our limits and achieve more than we otherwise would, to try even those borderline cases where a less confident person would bow out. A bit of excess confidence doesn't hurt; a little bit of above-average sensation can go a long way toward our psychological well-being and even our effectiveness at problem solving. When we're more confident, we take on tougher problems than we otherwise might. We push ourselves beyond our comfort zone.

But there can be such a thing as being too certain of yourself: overconfidence, when confidence trumps accuracy. We become more confident of our abilities, or of our abilities as compared with others', than we should be, given the circumstances and the reality. The illusion of validity grows ever stronger, the temptation to do things as you do ever more tempting. This surplus of belief in ourselves can lead to unpleasant results—like being so incredibly wrong about a case when you are usually so incredibly right, thinking a daughter is a husband, or a loving mother, a blackmailed wife.

It happens to the best of us. In fact, as I've hinted at already, it happens *more* to the best of us. Studies have shown that with experience, overconfidence *increases* instead of decreases. The more you know and the better you are in reality, the more likely you are to overestimate your own ability—and underestimate the force of events beyond your control. In one study, CEOs were shown to become more overconfident as they gained mergers and acquisitions experience: their estimates of a deal's value become overly optimistic (something not seen in earlier deals). In another, in contributions to pension plans, overconfidence correlated with age and education, such that the most overconfident contributors were highly educated males nearing retirement. In research from the University of Vienna, individuals were found to be, in general, *not* overconfident in their risky asset trades in an experimental market—until, that is, they obtained significant experience with the market in question. Then levels of overconfidence rose apace. What's more, analysts who have been more accurate at predicting earnings in the prior four quarters have been shown to be less accurate in subsequent earnings predictions, and professional traders tend to have a higher degree of overconfidence than students. In fact, one of the best predictors of overconfidence is power, which tends to come with time and experience.

Success breeds overconfidence like nothing else. When we are nearly always right, how far is it to saying that we'll always be right? Holmes has every reason to be confident. He is almost invariably correct, almost invariably better than anyone else at almost everything, be it thinking, solving crimes, playing the violin, or wrestling. And so, he should rightly fall victim to overconfidence often. His saving grace, however, or what is usually his saving grace, is precisely what we identified in the last section: that he knows the pitfalls of his mental stature and fights to avoid them by following his strict thought guidelines, realizing that he needs to always keep learning.

For those of us who live *off* the page, overconfidence remains a tricky thing. If we let our guard down for just a moment, as Holmes does here, it will get us.

Overconfidence causes blindness, and blindness in turn causes blunders. We become so enamored of our own skill that we discredit information that experience would otherwise tell us shouldn't be discredited—even information as glaring as Watson telling us that our theories are "all surmise," as he does in this case—and we proceed as before. We are blinded for a moment to everything we know about not theorizing before the facts, not getting ahead of ourselves, prying deeper and observing more carefully, and we get carried away by the simplicity of our intuition.

Overconfidence replaces dynamic, active investigation with passive assumptions about our ability or the seeming familiarity of our situation. It shifts our assessment of what leads to success from the conditional to the essential. *I am skilled enough that I can beat the environment as easily as I have been doing. Everything is due to my ability, nothing due to the fact that the surroundings just so happened to provide a good background for my skill to shine. And so I will not adjust my behavior.*

Holmes fails to consider the possibility of unknown actors in the drama or unknown elements in Mrs. Munro's biography. He also does not consider the possibility of disguise (something of a blind spot for the detective. If you remember, he, with equal confidence, does not take it into account in the case of Silver Blaze; nor does he do so in "The Man with the Twisted Lip"). Had Holmes had the same benefit of rereading his own exploits as we do, he may have learned that he was prone to this type of error.

Many studies have shown this process in action. In one classic demonstration, clinical psychologists were asked to give confidence judgments on a personality profile. They were given a case report in four parts, based on an actual clinical case, and asked after each part to answer a series of questions about the patient's personality, such as his behavioral patterns, interests, and typical reactions to life events. They were also asked to rate their confidence in their responses. With each section, background information about the case increased.

As the psychologists learned more, their confidence rose—but accuracy remained at a plateau. Indeed, all but two of the clinicians became overconfident (in other words, their confidence outweighed their accuracy), and while the mean level of confidence rose from 33 percent at the first stage to 53 percent by the last, the accuracy hovered at under 28 percent (where 20 percent was chance, given the question setup).

Overconfidence is often directly connected to this kind of underperformance—and at times, to grave errors in judgment. (Imagine a clinician in a nonexperimental setting trusting too much in his however inaccurate judgment. Is he likely to seek a second opinion or advise his patient to do so?) Overconfident individuals trust too much in their own ability, dismiss too easily the influences that they cannot control, and underestimate others—all of which leads to them doing much worse than they otherwise would, be it blundering in solving a crime or missing a diagnosis.

The sequence can be observed over and over, even outside of experimental settings, when real money, careers, and personal outcomes are at stake. Overconfident traders have been shown to perform worse than their less confident peers. They trade more and suffer lower returns. Overconfident CEOs have been shown to overvalue their companies and delay IPOs, with negative effects. They are also more likely to conduct mergers in general, and unfavorable mergers in particular. Overconfident managers have been shown to hurt their firms' returns. And overconfident detectives have been shown to blemish their otherwise pristine record through an excess of self-congratulation.

Something about success has a tendency to bring about an end to that very essential process of constant, never-ending education—unless the tendency is actively resisted, and then resisted yet again. There's nothing quite like victory to cause us to stop questioning and challenging ourselves in the way that is essential for Holmesian thinking.

Learning to Spot the Signs of Overconfidence

Perhaps the best remedy for overconfidence is knowing when it is most likely to strike. Holmes, for one, knows how liable past success and experience are to cause a blunder in thought. It is precisely this knowledge

that lets him lay his master trap for the villain at the heart of the trage-
dies in *The Hound of the Baskervilles*. When the suspect learns that Sherlock Holmes has arrived at the scene, Watson worries that the knowledge
will prove to make his capture all the more difficult: "I am sorry that he
has seen you," he tells Holmes. But Holmes is not so sure that it's a bad
thing. "And so was I at first," he responds. But now he realizes that the
knowledge, "may drive him to desperate measures at once. Like most clever
criminals, he may be too confident in his own cleverness and imagine
that he has completely deceived us."

Holmes knows that the successful criminal is likely to fall victim to
his very success. He knows to watch out for the red flag of cleverness that
thinks itself too clever, thereby underestimating its opponents while
overestimating its own strength. And he uses that knowledge in his capture of the villain on multiple occasions—not just at Baskerville Hall.

Spotting overconfidence, or the elements that lead to it, in others is
one thing; identifying it in ourselves is something else entirely, and far
more difficult. Hence Holmes's Norbury blunders. Luckily for us, however, psychologists have made excellent headway in identifying where
overconfidence most often lies in wait.

Four sets of circumstances tend to predominate. First, overconfidence
is most common when facing difficulty: for instance, when we have to
make a judgment on a case where there's no way of knowing all the facts.
This is called the hard-easy effect. We tend to be *under*confident on easy
problems and *over*confident on difficult ones. That means that we underestimate our ability to do well when all signs point to success, and we
overestimate it when the signs become much less favorable, failing to adjust enough for the change in external circumstances. For instance, in
something known as the choice-50 (C50) task, individuals must choose
between two alternatives and then state how confident they are in their
choice, between 0.5 and 1. Repeatedly, researchers have found that as the
difficulty of the judgment increases, the mismatch between confidence
and accuracy (i.e., overconfidence) increases dramatically.

One domain where the hard-easy effect is prevalent is in the making
of future predictions—a task that is nothing if not difficult (it is, as a matter of fact, impossible). The impossibility, however, doesn't stop people

from trying, and from becoming a bit too confident in their predictions based on their own perceptions and experience. Consider the stock market. It's impossible to actually predict the movement of a particular stock. Sure, you might have experience and even expertise—but you are nevertheless trying to predict the future. Is it such a surprise, then, that the same people who at times have outsized success also have outsized failures? The more successful you are, the more likely you are to attribute everything to your ability—and not to the luck of the draw, which, in all future predictions, is an essential part of the equation. (It's true of all gambling and betting, really, but the stock market makes it somewhat easier to think you have an inside, experiential edge.)

Second, overconfidence increases with familiarity. If I'm doing something for the first time, I will likely be cautious. But if I do it many times over, I am increasingly likely to trust in my ability and become complacent, even if the landscape should change (overconfident drivers, anyone?). And when we are dealing with familiar tasks, we feel somehow safer, thinking that we don't have the same need for caution as we would when trying something new or that we haven't seen before. In a classic example, Ellen Langer found that people were more likely to succumb to the illusion of control (a side of overconfidence whereby you think you control the environment to a greater extent than you actually do) if they played a lottery that was familiar versus one that was unknown.

It's like the habit formation that we've been talking about. Each time we repeat something, we become better acquainted with it and our actions become more and more automatic, so we are less likely to put adequate thought or consideration into what we're doing. Holmes isn't likely to pull a Yellow Face–style mess-up on his early cases; it's telling that the story takes place later in his career, and that it seems to resemble a more traditional blackmail case, the likes of which he has experienced many times before. And Holmes knows well the danger of familiarity, at least when it comes to others. In "The Adventure of the Veiled Lodger," he describes the experience of a couple who had fed a lion for too long. "It was deposed at the inquest that there has been some signs that the lion was dangerous, but, as usual, familiarity begat contempt, and no notice was taken of the fact." All Holmes has to do is apply that logic to himself.

Third, overconfidence increases with information. If I know more about something, I am more likely to think I can handle it, even if the additional information doesn't actually add to my knowledge in a significant way. This is the exact effect we observed earlier in the chapter with the clinicians who were making judgments on a case: the more information they had about the patient's background, the more confident they were in the accuracy of the diagnosis, yet the less warranted was that confidence. As for Holmes, he has detail upon detail when he travels to Norbury. But all the details are filtered through the viewpoint of Mr. Munro, who is himself unaware of the most important ones. And yet everything seems so incredibly plausible. Holmes's theory certainly covers all of the facts—the known facts, that is. But Holmes doesn't calibrate for the possibility that, despite the magnitude of the information, it continues to be *selective* information. He lets the sheer amount overwhelm what should be a note of caution: that he still knows nothing from the main actor who could provide the most meaningful information, Mrs. Munro. As ever, quantity does not equal quality.

Finally, overconfidence increases with action. As we actively engage, we become more confident in what we are doing. In another classic study, Langer found that individuals who flipped a coin themselves, in contrast to watching someone else flip it, were more confident in being able to predict heads or tails accurately, even though, objectively, the probabilities remained unchanged. Furthermore, individuals who chose their own lottery ticket were more confident in a lucky outcome than they were if a lottery ticket was chosen for them. And in the real world, the effects are just as pronounced. Let's take the case of traders once again. The more they trade, the more confident they tend to become in their ability to make good trades. As a result, they often overtrade, and in so doing undermine their prior performance.

But forewarned is forearmed. An awareness of these elements can help you counteract them. It all goes back to the message at the beginning of the chapter: we must continue to learn. The best thing you can do is to acknowledge that you, too, will inevitably stumble, be it from stagnation or overconfidence, its closely related near opposite (I say *near* because overconfidence creates the illusion of movement, as opposed to habitual stagnation, but that movement isn't necessarily taking you any-where), and to keep on learning.

As "The Yellow Face" draws to a close, Holmes has one final message for his companion. "Watson, if it should ever strike you that I am getting a little overconfident in my powers, or giving less pains to a case than it deserves, kindly whisper 'Norbury' in my ear, and I shall be infinitely obliged to you." Holmes was right about one thing: he shouldn't have missed the case for worlds. Even the best of us—especially the best of us—need a reminder of our fallibility and ability to deceive ourselves into a very confident blunder.

Now for the Good News:
It's Never Too Late to Keep Learning, Even After You've Stopped.

We opened the chapter with "The Red Circle," Holmes's triumph of never-ending education. The year of that feat of undying curiosity and ever-present desire to continue to challenge the mind with new, more dif-ficult cases and ideas? 1902.[4] As for the year of "The Yellow Face," when victory of confidence over the very education Holmes urges befell the great detective? 1888. I raise this chronology to point out one somewhat obvious and yet absolutely central element of the human mind: we never stop learning. The Holmes that took the case of a mysterious lodger and ended up embroiled in a saga of secret societies and international crime rings (for that is the meaning of Red Circle: a secret Italian crime syndi-cate with many evil deeds to its name) is no longer the same Holmes who made such seemingly careless errors in "The Yellow Face."

Holmes may have his Norburys. But he has chosen to learn from them and make himself a better thinker in the process, ever perfecting a mind that already seems sharp beyond anything else. We, too, never stop learning, whether we know it or not. At the time of "The Red Circle," Holmes was forty-eight years old. By traditional standards, we might have thought him incapable of any profound change by that point in life,

---

[4] All cases and Holmes's life chronology are taken from Leslie Klinger's *The New Annotated Sherlock Holmes* (NY: W. W. Norton, 2004).

at least on the fundamental level of the brain. Until recently, the twenties were considered the final decade during which substantial neural changes could take place, the point where our wiring is basically complete. But new evidence points to an altogether different reality. Not only can we keep learning but our brains' very structure can change and develop in more complex ways for far longer, even into old age.

In one study, adults were taught to juggle three balls over a three-month period. Their brains, along with those of matched non-juggling adults who received no training, were scanned at three points in time: before the training began, at a point when they reached juggling proficiency (i.e., could sustain the routine for at least sixty seconds), and three months after the proficiency point, during which time they were asked to stop juggling altogether. At first there were no differences in gray matter between jugglers and non-jugglers. By the time the jugglers had reached proficiency, however, a marked change was apparent: their gray matter had increased bilaterally (i.e., in both hemispheres) in the mid-temporal area and the left posterior intraparietal sulcus, areas associated with the processing and retention of complex visual-motion information. Not only were the jugglers learning, but so were their brains—and learning at a more fundamental level than previously thought possible.

What's more, these neural changes can happen far more rapidly than we've ever realized. When researchers taught a group of adults to distinguish newly defined and named categories for two colors, green and blue, over a period of two hours (they took four colors that could be told apart visually but not lexically and assigned arbitrary names to each one), they observed an increase in gray-matter volume in the region of the visual cortex that is known to mediate color vision, V2/3. So in just two hours the brain was already showing itself receptive to new inputs and training, at a deep, structural level.

Even something that has been traditionally seen as the purview of the young—the ability to learn new languages—continues to change the landscape of the brain late into life. When a group of adults took a nine-month intensive course in modern standard Chinese, their brains' white matter reorganized progressively (as measured monthly) in the left hemisphere language areas and their right hemisphere counterparts—as well as in the *genu* (anterior end) of the corpus collosum, that network of neural fibers that connects the two hemispheres, which we encountered in the discussion of split-brain patients.

And just think of the rewiring that takes place in extreme cases, when a person loses his vision or function in some limb or undergoes some other drastic change in the body. Entire areas of the brain become reassigned to novel functions, taking up the real estate of the lost faculty in intricate and innovative ways. Our brains are capable of learning feats that are nothing short of miraculous.

But there's more. It now seems clear that with application and practice even the elderly can reverse signs of cognitive decline *that has already occurred.* I place that emphasis out of pure excitement. How amazing to consider that even if we've been lazy all our lives, we can make a substantial difference and reverse damage that has already been done, if only we apply ourselves and remember Holmes's most enduring lesson.

There is, of course, a downside in all this. If our brains can keep learning—and keep changing as we learn—throughout our lives, so, too, can they keep unlearning. Consider this: in that juggling study, by the time of the third scan, the gray-matter expansion that had been so pronounced three months prior had decreased drastically. All of that training? It had started to unravel at every level, performance and neural. What does that mean? Our brains are learning whether we know it or not. If we are not strengthening connections, we are losing them.

Our education might stop, if we so choose. Our brains' never does. The brain will keep reacting to how we decide to use it. The difference is not whether or not we learn, but what and how we learn. We can learn to be passive, to stop, to, in effect, not learn, just as we can learn to be curious, to search, to keep educating ourselves about things that we didn't even know we needed to know. If we follow Holmes's advice, we teach our brains to be active. If we don't, if we're content, if we get to a certain point and decide that that point is good enough, we teach them the opposite.

SHERLOCK HOLMES FURTHER READING

"It's a police matter, Mr. Holmes!" "It is art for art's sake." from *His Last Bow*, "The Adventure of the Red Circle," p. 1272.

"Come at once if convenient." "As an institution I was like the violin, the shag tobacco, the old black pipe, the index books." from *The Memoirs of Sherlock Holmes*, "The Crooked Man," p. 138.

"There's blackmail in it, or I am much mistaken." from *The Memoirs of Sherlock Holmes*, "The Yellow Face," p. 30.

"Like most clever criminals, he may be too confident in his own clever-ness . . ." from *The Hound of the Baskervilles*, chapter 12: Death on the Moor, p. 121.

PART FOUR

THE SCIENCE AND ART OF SELF-KNOWLEDGE

CHAPTER SEVEN

The Dynamic Attic:
Putting It All Together

In the opening pages of *The Hound of the Baskervilles*, Watson enters the sitting room of 221B Baker Street to find a walking stick that has been left behind by a certain James Mortimer. When he takes the opportunity to try to put Holmes's methods into practice, seeing what he can deduce about the doctor from the appearance of the stick, he finds his thoughts interrupted by his friend.

"Well, Watson, what do you make of it?" Holmes asks.

Watson is shocked. Holmes had been sitting at the breakfast table, with his back turned. How could he have known what the doctor was doing or thinking? Surely, he must have eyes in the back of his head.

Not quite, says Holmes. "I have, at least, a well-polished, silver-plated coffee-pot in front of me. But tell me, Watson, what do you make of our visitor's stick?" he presses. "Let me hear you reconstruct the man by an examination of it."

Watson gamely takes up the challenge, trying his best to mirror his companion's usual approach. "I think that Dr. Mortimer is a successful, elderly medical man, well-esteemed, since those who know him give him this mark of their appreciation," he begins. "I also think that the probability is in favour of his being a country practitioner who does a great deal of his visiting on foot."

The first part initially sounds reasonable enough. But why does Watson deduce the second? "Because this stick, though originally a very handsome one, has been so knocked about that I can hardly imagine a town practitioner carrying it," he says.

Holmes is pleased. "Perfectly sound!" he exclaims. And what else?

"And then again, there is the 'friends of the C.C.H.,'" Watson notes the inscription on the stick. "I should guess that to be the Something

Hunt, the local hunt to whose members he has possibly given some surgical assistance," he continues, "and which has made him a small presentation in return."

"Really, Watson, you excel yourself," Holmes responds. He then goes on to praise Watson as a "conductor of light" and a stimulator of genius, ending his paean with the words, "I must confess, my dear fellow, that I am very much in your debt."

Has Watson finally learned the trick? Has he mastered Holmes's reasoning process? Well, for at least a moment he basks in the compliment. Until, that is, Holmes picks up the stick himself and comments that there are indeed "one or two indications" that can furnish the basis for deduction.

"Has anything escaped me?" Watson asks with admitted self-importance. "I trust that there is nothing of consequence which I have overlooked?"

Not exactly. "I am afraid, my dear Watson, that most of your conclusions were erroneous," Holmes says. "When I said that you stimulated me I meant, to be frank, that in noting your fallacies I was occasionally guided towards the truth. Not that you are entirely wrong in this instance. The man is certainly a country practitioner. And he walks a good deal."

Watson takes that to mean that he had, in point of fact, been right. Well, only insofar as he got those details accurately. But is he still right if he fails to see the bigger picture?

Not according to Holmes. He suggests, for instance, that C.C.H. is much more likely to refer to Charing Cross Hospital than to any local hunt, and that from there stem multiple inferences. What may those be, wonders Watson?

"Do none suggest themselves?" Holmes asks. "You know my methods. Apply them!"

And with that famed interjection, that challenge, if you will, Holmes embarks on his own logical tour de force, which ends with the arrival of Dr. Mortimer himself, followed closely by the curly-haired spaniel whose existence the detective has just deduced.

. . .

This little repartee brings together all of the elements of the scientific approach to thought that we've spent this book exploring and serves as a near-ideal jumping-off point for discussing how to bring the thought process together as a whole—and how that coming together may fall short. That walking stick illustrates both how to think properly and how one can fail to do so. It presents that crucial line between theory and practice, between the knowledge of how we're to think and the practice of actually doing so.

Watson has observed Holmes at work many a time, and yet when it comes to applying the process himself, he remains unsuccessful. Why? And how can we do him one better?

1. Know Yourself—And Your Environment

We begin, as always, with the basics. What are we ourselves bringing to a situation? How do we assess the scene even before we begin the observational process?

To Watson, the question at hand begins with the walking stick: "a fine, thick piece of wood, bulbous-headed, of the sort which is known as a 'Penang lawyer,' " which is "just such a stick as the old-fashioned family practitioner used to carry—dignified, solid, and reassuring." That first bit is just fine, a description of the stick's outward qualities. But take a close look at the second part. Is that true observation, or is it more like inference?

Hardly has Watson started to describe the stick and already his personal biases are flooding his perception, his own experience and history and views framing his thoughts without his realizing it. The stick is no longer just a stick. It is the stick of the old-fashioned family practitioner, with all the characteristics that follow from that connection. The instantaneously conjured image of the family doctor will color every judgment that Watson makes from this point forward—and he will have no idea that it is doing so. In fact, he will even fail to consider that C.C.H. might stand for a prominent hospital, something that he as a doctor himself should be well aware of, if only he'd not gone off on the country doctor tangent and failed to consider it entirely.

This is the frame, or the subconscious prime, in all its glory. And who knows what other biases, stereotypes, and the like will be rustled up out of the corners of Watson's brain attic along with it? Certainly not he. But we can know one thing. Any heuristics—or rules of thumb, as you'll recall—that will affect his eventual judgment will likely have their root in this initial, thoughtless assessment.

Holmes, on the other hand, realizes that there is always a step that comes *before* you begin to work your mind to its full potential. Unlike Watson, he doesn't begin to observe without quite being aware of it, but rather takes hold of the process from the very beginning—and starting well before the stick itself. He takes in the whole situation, doctor and stick and all, long before he starts to make detailed observations about the object of interest itself. And to do it, he does something far more prosaic than Watson could ever suppose: he looks in a polished silver coffeepot. He doesn't need to use his deductional powers where he has use of a reflective surface; why waste them needlessly?

So, too, must we always look around us to see if there's a ready-and-waiting mirror, before plunging in without a second thought—and then use it to take stock of the entire situation instead of letting the mind thoughtlessly get ahead of itself and begin grabbing who knows what out of our attic without our full knowledge and control.

Evaluating our environment means different things, depending on the choices we are making. For Holmes, it was observing the room, Watson's actions, and the easily available coffeepot. Whatever it is, we can rest assured that it will require a pause before the dive. We can't forget to look at our surroundings before launching into action—or even into the Holmesian thought process. For, after all, pausing and reflecting is the first step to that process. It's point zero of observation. Before we begin to gather detail, we need to know what detail, if any, we'll be gathering.

Remember: specific, mindful motivation matters. It matters a great deal. We have to frame our goals ahead of time. Let them inform how we proceed. Let them inform how we allocate our precious cognitive resources. We have to think them through, write them down, to make sure they are as clear-cut as they can possibly be. Holmes doesn't need to take notes, to be sure, but most of us certainly do, at least for the truly important choices. It will help clarify the important points before we embark on our journey of thought: *What do I want to accomplish? And what does that mean for my future thought process?* Not looking necessarily means not finding. And to find, we first need to know where to look.

2. Observe—Carefully and Thoughtfully

When Watson looks at the stick, he notes its size and heft. He also remarks the beat-up bottom—a sign of frequent walking in terrain that is less than hospitable. Finally, he looks to the inscription, *C.C.H.*, and with that concludes his observations, confident as ever that nothing has escaped his notice.

Holmes, on the other hand, is not so sure. First off, he does not limit his observation to the stick as physical object; after all, the original goal, the frame set in the first step of the process, was to learn about the man who owned it. "It is only an absent-minded one who leaves his stick and not his visiting-card after waiting an hour in your room," he tells Watson. But of course: the stick was left behind. Watson knows that, naturally—and yet he fails to *know* it.

What's more, the stick creates its own context, its own version of the owner's history, if you will, by virtue of the inscription. While Watson reads the *C.C.H.* only in light of his unconscious preconceptions of the country practitioner, Holmes realizes that it must be observed on its own terms, without any prior assumptions, and that in that light, it tells its own story. Why would a doctor receive a stick as a gift? Or, as Holmes puts it, "On what occasion would it be most probable that such a presentation would be made? When would his friends unite to give him a pledge of their good will?" That is the point of departure suggested by a true observation of the inscription, not a biased one, and that point suggests a background story that can be reached through careful deduction. The context is an integral part of the situation, not a take-it-or-leave-it accessory.

As for the stick itself, here, too, the good doctor has not been as careful in his observations as he should have been. First off, he merely glances at it, whereas Holmes "examined it for a few minutes with his naked eyes. Then with an expression of interest he laid down his cigarette, and, carrying the cane to the window, he looked over it again with a convex lens." Closer scrutiny, from multiple angles and multiple approaches. Not as fast as the Watson method, to be sure, but much more thorough. And while it may well be true that such care will not be rewarded with any new details, you can never know in advance, so if you are to truly observe, you can never afford to forego it. (Though, of course, our own window and convex lens may be metaphorical, they nevertheless imply a degree of closer scrutiny, of scrupulousness and sheer time spent in contemplation of the problem.)

Watson notes the stick's size and the worn-down bottom, true. But he fails to see that there are teeth marks plainly visible on its middle. Teeth marks on a stick? It's hardly a leap of faith to take that observation as implying the existence of a dog who has carried the stick, and carried it often, behind his master (as Holmes, in fact, does). That, too, is part of the observation, part of the full story of Dr. Mortimer. What's more, as Holmes points out to his friend, the size of the dog's jaw is evident from the space between the marks, making it possible to envision just what type of dog it might have been. That, of course, would be jumping ahead to deduction—but it wouldn't be possible at all without recognizing the necessary details and mentally noting their potential significance for your overall goal.

3. Imagine—Remembering to Claim the Space You May Not Think You Need

After observation comes that creative space, that time to reflect and explore the ins and outs of your attic called imagination. It's that break of the mind, that three-pipe problem, that violin interlude or opera or concerto or trip to the art museum, that walk, that shower, that who knows what that forces you to take a step back from the immediacy of the situation before you once more move forward.

We need to give Watson some credit here. He doesn't exactly have time to take a break, as Holmes puts him on the spot, challenging him to apply the detective's methods to inferring what he can about the implications of C.C.H. standing for Charing Cross Hospital instead of for Something Hunt. Watson can hardly be expected to break out the cigarettes or brandy.

And yet Watson could do something a little less extreme but far more appropriate to a problem of far lesser magnitude than solving a full crime. After all, not everything is a three-pipe problem. It may be enough to take a more metaphorical step back. To distance yourself mentally, to pause and reflect and reconfigure and reintegrate in a much shorter time frame.

But Watson does no such thing. He doesn't even give himself time to think after Holmes prompts him to do so, saying that he can only draw "the obvious conclusions" but can't see anything further.

Contrast the approach that Watson and Holmes take. Watson goes right to it: from observation of the heft and shape of the stick to image of old-fashioned practitioner, from *C.C.H.* to Something Hunt, from worn-down iron ferrule to country practitioner, from Charing Cross to a move from town to country, and nothing more besides. Holmes, on the other hand, spends quite a bit more time in between his observations and his conclusions. Recall that first, he listens to Watson; next, he examines the stick; then, he once more speaks with Watson; and finally, when he begins to list his own conclusions, he does not do so all at once. Rather, he asks himself questions, questions that suggest a number of answers, before settling on a single possibility. He looks at different permutations—could Dr. Mortimer have been in a well-established London practice? A house surgeon? A house physician? A senior student?—and then considers which would be more likely in light of all of the other observations. He doesn't deduce. Rather, he reflects and he plays around with options. He questions and he considers. Only *after* will he start to form his conclusions.

4. Deduce—Only from What You've Observed, and Nothing More

From a walking stick to a "successful, elderly medical man, well-esteemed," a "country practitioner who does a lot of his visiting on foot" and who has "given some surgical assistance" to a local hunt (for which he has received said stick), if you're Watson. And from that same stick to a former Charing Cross Hospital "house-surgeon or house-physician," a "young fellow under thirty, amiable, unambitious, absent-minded, and the possessor of a favourite dog"—nay, a curly-haired spaniel—who received the stick on the occasion of the change from Charing Cross to the country, if you're Holmes. Same starting point, altogether different deductions (with the sole intersection of a country practitioner who walks a great deal). How do two people come out so differently when faced with an identical problem?

Watson has made two correct deductions: that the stick belongs to a country practitioner and that that practitioner does much of his visiting on foot. But why elderly and well esteemed? Whence came this picture of the conscientious and dedicated family practitioner? Not from any actual observation. It came instead from a fabrication of Watson's mind, of his immediate reaction that the stick was just such "as the old-fashioned family practitioner used to carry—dignified, solid, and reassuring."

The stick itself is no such thing, other than solid. It is just an object that carries certain signs. But to Watson, it at once has a story. It has brought up memories that have little bearing on the case at hand and instead are stray pieces of attic furniture that have become activated by virtue of some associative memory processes of which Watson himself is hardly aware. Ditto the local hunt. So focused has Watson become on his imagined solid and dignified country practitioner that it seems only logical to him that the walking stick was the gift of a hunt, to whose members Dr. Mortimer has, naturally, given some surgical assistance. Watson doesn't actually have any solid, logical steps to show for these deductions. They stem from his selective focus and the doctor that exists in his imagination. As a reassuring and elderly family man, Dr. Mortimer would naturally be both a member of a local hunt and ever ready to give assistance. Surgical? But of course. Someone of such stature and refinement must clearly be a surgical man.

Watson fails to note entirely the M.R.C.S. appended to Mortimer's name (something that the man himself will later point out in correcting

Holmes when the latter addresses him as Doctor: "Mister, sir, Mister—a humble M.R.C.S.")—an addition that belies the stature Mortimer has assumed in Watson's hyperactive mind. And he makes no note, as we've already discussed, of the sheer fact of the stick having been left in the visiting room—minus so much as a visiting card. His memory in this instance is as mindlessly selective as his attention—after all, he did read the M.R.C.S. when he first looked at the stick; it was just overshadowed completely by the details his mind then supplied of its own accord based on the nature of the stick itself. And he did recognize at the very beginning that the stick's owner had left it behind on the prior evening, but that, too, slipped his mind as an observation or fact worthy of note.

Holmes's version, in contrast, comes from an entirely different thought process, one that is fully aware of itself and of its information, that seeks to incorporate all evidence and not just selective bits, and to use that evidence as a whole, rather than focusing on some parts but not others, coloring some more brightly, and others in a paler hue.

First, the man's age. "You will observe," he tells Watson, after having convinced the doctor that the most likely meaning of C.C.H. is Charing Cross Hospital and not Something Hunt (after all, we are talking about a *doctor*; isn't it most logical that he would receive a presentation from a hospital and not a hunt? Which of the two *Hs* is the more likely, given the objective information and not any subjective version thereof?), "that he could not have been on the staff of the hospital, since only a man well-established in a London practice could hold such a position, and such a one would not drift to the country." (We know, of course, that drift to the country the man did, based on the indications of the stick, the very ones that Watson so eagerly noted and grasped.) Fair enough. Someone so well established as to be a staff member would hardly be expected to up and leave—unless, of course, there were some unforeseen circumstances. But there are no such circumstances that one could grasp from the evidence of the stick, so that is not an explanation to be considered from the available evidence (indeed, considering it would entail the precise fallacy that Watson commits in creating his version of the doctor, a story told by the mind and not based in objective observation).

Who, then? Holmes reasons it out: "If he was in the hospital and yet not on the staff he could only have been a house-surgeon or a house-physician—little more than a senior student. And he left five years ago—the date is on the stick." Hence, "a young fellow under thirty" to Watson's middle-aged practitioner. Note also that while Holmes is certain about the age—after all, he has exhausted all options of his former position, until only one reasonable age alternative remains (remember: "It may well be that several explanations remain, in which case one tries to test after test until one or other of them has a convincing amount of support")—he does *not* go as far as Watson in necessitating that the man in question be a surgeon. He may just as well be a physician. There is zero evidence to point in either direction, and Holmes does not deduce past where the evidence leads. That would be just as fallacious as not deducing far enough.

What of the man's personality? "As to the adjectives," says Holmes, "I said, if I remember right, amiable, unambitious, and absent-minded." (He does remember right.) How could he have possibly deduced these characteristics? Not, it turns out, in the mindless fashion that Watson deduced his own set of attributes. "It is my experience," says Holmes, "that it is only an amiable man in this world who receives testimonials, only an unambitious one who abandons a London career for the country, and only an absent-minded one who leaves his stick and not his visiting-card after waiting an hour in your room." Each trait emerges directly from one of the observations (filtered through the time and space of imagination, even if only for a span of some minutes) that Holmes has made earlier.

Objective fact, to a consideration of multiple possibilities, to a narrowing of the most likely ones. No extraneous details, no holes filled in by an all too willing imagination. Scientific deduction at its best.

Finally, why does Holmes give Dr. Mortimer a dog, and a very specific one at that? We've already discussed the teeth marks that Watson has missed. But the marks—or rather, the distance between them—are quite specific, "too broad in my opinion for a terrier and not broad enough for a mastiff." Holmes may well have gotten to a curly-haired spaniel on his own, following that logical train, but he has no opportunity to do so, as the dog in question appears at that moment alongside its owner. And there, the deductive trail comes to an end. But wasn't it a clear one as far as it went? Didn't it make you want to say, *Elementary? How could I not have seen that myself?* That is exactly what deduction at its best, of course, is meant to do.

5. Learn—From Your Failures Just as You Do from Your Successes

In observing Watson's fallacies in this particular instance, Holmes learns ever more about the pitfalls of the thought process, those moments when it is easy to go astray—and precisely in which direction the false path usually lies. From this encounter, he will take away the power of stereotype activation and the overwhelming influence an improper initial frame can have on the inferences that follow, as well as the error that is introduced when one fails to consider every observation and focuses instead on the most salient, recent, or otherwise accessible ones. Not that he doesn't know both of these things already, but each time serves as a reminder, a reinforcement, a new manifestation in a different context that ensures that his knowledge never goes stale.

And if Watson is paying close attention, he should take away much the same things, learning from Holmes's corrections to identify those moments where he went wrong and to learn how better to go right the next time around. Alas, he chooses the other route, focusing instead on Holmes's statement that he is not "entirely wrong in this instance. The man is certainly a country practitioner. And he walks a good deal." Instead of trying to see why it was precisely that he got these two details right and the rest altogether wrong, Watson says, "Then I was right," forsaking the opportunity to learn, and instead focusing once more only selectively on the available observations.

Education is all well and good, but it needs to be taken from the level of theory to that of practice, over and over and over—lest it begin to gather dust and let out that stale, rank smell of the attic whose door has remained unopened for years.

Any time we get the urge to take it easy, we'd do well to bring to mind the image of the rusted razor blade from *The Valley of Fear*: "A long series of sterile weeks lay behind us, and here at last was a fitting object for those remarkable powers which, like all special gifts, become irksome to their owner when they are not in use. That razor brain blunted and rusted with inaction." Picture that rusted, blunted razor, the yucky orange specks peeling off, the dirt and decay so palpable that you don't even want to reach out to remove it from its place of neglect, and remember that even when everything seems wonderful and there are no major choices to be made or thoughts to be thought, the blade has to remain in use. Exercising our minds even on the unimportant things will help keep them sharp for the important ones.

Time to Keep a Diary

Let's take a quick break from Mr. Mortimer. A good friend of mine—I'll call her Amy—has long been a migraine sufferer. Everything will be going just fine when out of the blue, it hits her. Once, she thought she was dying, another time, that she'd gotten the terrible Norovirus that had been going around. It took some years for her to learn to discern the first signs and run for the nearest dark room and a nice dose of Imitrex *before* the I'm-about-to-die/I-have-a-horrible-stomach-flu panic set in. But eventually, she could more or less manage. Except when the migraines struck several times a week, putting her behind work, writing, and everything else in a steady stream of pain. Or when they came at those inopportune times when she had neither a dark, quiet room nor medicine to fall back on. She soldiered on.

A year or so ago, Amy switched primary care doctors. During the usual getting-to-know-you chat, she complained, as always, about her migraines. But instead of nodding sympathetically and prescribing more Imitrex, as every doctor before her had done, this particular physician asked her a question. Had Amy ever kept a migraine diary?

Amy was confused. Was she supposed to write from the migraine's point of view? Try to see through the pain and describe her symptoms for posterity? No. It was much simpler. The doctor gave her a stack of pre-printed sheets, with fields like Time Started/Ended, Warning Signs, Hours of Sleep, what she'd eaten that day, and the lot. Each time Amy had a migraine, she was to fill it in retroactively, as best she could. And she was to keep doing it until she had a dozen or so entries.

Amy called me afterward to tell me just what she thought of the new doctor's approach: the whole exercise was rather absurd. She knew what caused her migraines, she told me confidently. It was stress and changes of weather. But she said she'd give it a shot, if only for a laugh and despite her reservations. I laughed right along with her.

I wouldn't be telling the story now if the results didn't shock us both. Did caffeine ever cause migraines? the doctor had asked Amy in their initial conversation. Alcohol? Amy had shaken her head knowingly. Absolutely not. No connection whatsoever. Except that's not the story the migraine diary told. Strong black tea, especially later in the day, was almost always on the list of what she'd eaten before an attack. More than a glass of wine, also a frequent culprit. Hours of sleep? Surely that wasn't important. But there it was. The number of hours listed on those days when she found it hard to move tended to be far below the usual amount. Cheese (cheese? seriously?), also on the list. And, yes, she had been right, too. Stress and changes in weather were surefire triggers.

Only, Amy hadn't been right entirely. She had been like Watson, insisting that she'd been correct, when she'd been correct only "to that extent." She'd just never taken notice of anything else, so salient were those two factors. And she certainly never drew the connections that were, in retrospect, all too apparent.

Knowing is only part of the battle, of course. Amy still gets migraines more often than she would like. But at the very least, she can control some of the trigger factors much better than she ever could before. And she can spot symptoms earlier, too, especially if she's knowingly done something she shouldn't, like have some wine *and* cheese . . . on a rainy day. Then she can sometimes sneak in the medicine before the headache sets in for good, and at least for the moment she has it beat.

Not everyone suffers from migraines. But everyone makes choices and decisions, thinks through problems and dilemmas, on a daily basis. So here's what I recommend to speed up our learning and help us integrate all of those steps that Holmes has so graciously shown us: we should keep a decision diary. And I don't mean metaphorically. I mean actually, physically, writing things down, just as Amy had to do with her migraines and triggers.

When we make a choice, solve a problem, come to a decision, we can record the process in a single place. We can put here a list of our observations, to make sure we remember them when the time comes; we can include, too, our thoughts, our inferences, our potential lines of inquiry, things that intrigued us. But we can even take it a step further. Record what we ended up doing. Whether we had any doubts or reservations or considered other options (and in all cases, we'd do well to be specific and say what those were). And then, we can revisit each entry to write down how it went. *Was I happy? Did I wish I'd done something differently? Is there anything that is clear to me in retrospect that wasn't before?*

For those choices for which we haven't written any observations or made any lists, we can still try our best to put down what was going through our mind at the time. *What was I considering? What was I basing my decision on? What was I feeling in the moment? What was the context (was I stressed? emotional? lazy? was it a regular day or not? what, if anything, stood out?)? Who else, if anyone, was involved? What were the stakes? What was my goal, my initial motivation? Did I accomplish what I'd set out to do? Did something distract me?* In other words, we should try to capture as much as possible of our thought process and its result.

And then, when we've gathered a dozen (or more) entries or so, we can start to read back. In one sitting, we can look through it all. All of those thoughts on all of those unrelated issues, from beginning to end. Chances are that we'll see the exact same thing Amy did when she reread her migraine entries: that we make the same habitual mistakes, that we think in the same habitual ways, that we're prey to the same contextual cues over and over. And that we've never quite seen what those habitual patterns are—much as Holmes never realizes how little credit he gives to others when it comes to the power of disguise.

Indeed, writing things down that you think you know cold, keeping track of steps that you think need no tracking, can be an incredibly useful habit even for the most expert of experts. In 2006, a group of physicians released a groundbreaking study: they had managed to lower the rate of catheter-related bloodstream infections—a costly and potentially lethal phenomenon, estimated at about 80,000 cases (and up to 28,000 deaths) per year, at a cost of $45,000 per patient—in Michigan ICUs from a median rate of 2.7 infections in 1,000 patients to 0 in only three months. After sixteen and eighteen months, the mean rate per 1,000 had decreased from a baseline of 7.7 to 1.4 infections. How was this possible? Had the doctors discovered some new miracle technique?

Actually, they had done something so simple that many a physician rebelled at such a snub to their authority. They had instituted a mandatory checklist. The checklist had only five items, as simple as handwashing and making sure to clean a patient's skin prior to inserting the catheter. Surely, no one needed such elementary reminders. And yet— with the reminders in place, the rate of infection dropped precipitously, to almost zero. (Consider the natural implication: prior to the checklist, some of those obvious things weren't getting done, or weren't getting done regularly.)

Clearly, no matter how expert at something we become, we can forget the simplest of elements if we go through the motions of our tasks mindlessly, regardless of how motivated we may be to succeed. Anything that prompts a moment of mindful reflection, be it a checklist or something else entirely, can have profound influence on our ability to maintain the same high level of expertise and success that got us there to begin with.

Humans are remarkably adaptable. As I've emphasized over and over, our brains can wire and rewire for a long, long time. Cells that fire together wire together. And if they start firing in different combinations, with enough repetition, that wiring, too, will change.

The reason I keep focusing on the necessity of practice is that practice is the only thing that will allow us to apply Holmes's methodology in real life, in the situations that are far more charged emotionally than any thought experiment can ever lead you to believe. We need to train ourselves mentally for those emotional moments, for those times when the deck is stacked as high against us as it will ever be. It's easy to forget how quickly our minds grasp for familiar pathways when given little time to think or when otherwise pressured. But it's up to us to determine what those pathways will be.

It is most difficult to apply Holmes's logic in those moments that matter the most. And so, all we can do is practice, until our habits are such
that even the most severe stressors will bring out the very thought patterns that we've worked so hard to master.

SHERLOCK HOLMES FURTHER READING

*"You know my methods. Apply them!" "Well, Watson, what do you make
of it?"* from *The Hound of the Baskervilles,* chapter 1: Mr. Sherlock
Holmes, p. 5.
"If I take it up, I must understand every detail" from *His Last Bow,* "The
Adventure of the Red Circle," p. 1272.
"That razor brain blunted and rusted with inaction" from *The Valley of
Fear,* chapter 2: *Mr. Sherlock Holmes Discourses,* p. 11.

CHAPTER EIGHT

We're Only Human

On a morning in May 1920, Mr. Edward Gardner received a letter
from a friend. Inside were two small photographs. In one, a group
of what looked to be fairies were dancing on a stream bank while a little
girl looked on. In another, a winged creature (a gnome perhaps, he
thought) sat near another girl's beckoning hand.

Gardner was a theosophist, someone who believed that knowledge of
God may be achieved through spiritual ecstasy, direct intuition, or special individual relation (a popular fusion of Eastern ideas about reincar-
nation and the possibility of spirit travel). Fairies and gnomes seemed a
far cry from any reality he'd ever experienced outside of books, but where
another may have laughed and cast aside pictures and letter both, he was
willing to dig a little deeper. And so, he wrote back to the friend: Might
he be able to obtain the photo negatives?

When the plates arrived, Gardner promptly delivered them to a Mr.
Harold Snelling, photography expert extraordinaire. No fakery, it was
said, could get past Snelling's eye. As the summer drew on, Gardner

awaited the expert's verdict. Was it possible that the photographs were something more than a clever staging?

By the end of July, Gardner got his answer: "These two negatives," Snelling wrote, "are entirely genuine unfaked photographs of single exposure, open-air work, show movement in the fairy figures, and there is no trace whatever of studio work involving card or paper models, dark backgrounds, painted figures, etc. In my opinion, they are both straight untouched pictures."

Gardner was ecstatic. But not everyone was equally convinced. It seemed so altogether improbable. One man, however, heard enough to pursue the matter further: Sir Arthur Conan Doyle.

Conan Doyle was nothing if not meticulous. In that, at least, he took his creation's methodology to heart. And so, he asked for further validation, this time from an undisputed authority in photography, Kodak—who also happened to have manufactured the camera that had been used to take the picture.

Kodak refused to offer an official endorsement. The photographs were indeed single exposure, the experts stated, and showed no outward signs of being faked, but as for their genuineness, well, that would be taking it one step too far. The photographs *could* have been faked, even absent outward signs, and anyhow, fairies did not exist. Ergo, the pictures could not possibly be real.

Conan Doyle dismissed that last bit as faulty logic, a circular argument if ever there was one. The other statements, however, seemed sound enough. No signs of fakery. Single exposure. It certainly seemed convincing, especially when added to Snelling's endorsement. The only negative finding that Kodak had offered was pure conjecture—and who better than Holmes's creator to know to throw those out of consideration?

There remained, however, one final piece of evidence to verify: what about the girls depicted in the photographs? What evidence, be it supportive or damning, could they offer? Alas, Sir Arthur was leaving on a trip to Australia that would not be put off, and so, he asked Gardner to travel in his stead to the scene of the pictures, a small West Yorkshire town called Cottingley, to speak with the family in question.

In August 1920, Edward Gardner met Elsie Wright and her six-years-younger cousin, Frances Griffiths, for the first time. They'd taken the photographs, they told him, three years prior, when Elsie was sixteen and Frances ten. Their parents hadn't believed their tale of fairies by the stream, they said, and so they had decided to document it. The photographs were the result.

The girls, it seemed to Gardner, were humble and sincere. They were well-raised country girls, after all, and they could hardly have been after personal gain, refusing, as they did, all mention of payment for the pictures. They even asked that their names be withheld were the photographs to be made public. And though Mr. Wright (Elsie's father) remained skeptical and called the prints nothing more than a childish prank, Mr. Gardner was convinced that these photos were genuine: the fairies were real. These girls weren't lying. Upon his return to London, he sent a satisfied report to Conan Doyle. So far, everything seemed to be holding together.

Still, Conan Doyle decided that more proof was in order. Scientific experiments, after all, needed to be replicated if their results were to be held valid. So Gardner traveled once more to the country, this time with two cameras and two dozen specially marked plates that couldn't be substituted without drawing attention to the change. He left these with the girls with the instructions to capture the fairies again, preferably on a sunny day when the light was best.

He wasn't disappointed. In early fall, he received three more photographs. The fairies were there. The plates were the original ones he'd supplied. No evidence of tampering was found.

Arthur Conan Doyle was convinced. The experts agreed (though, of course, one without offering official endorsement). The replication had gone smoothly. The girls seemed genuine and trustworthy.

In December, the famed creator of Mr. Sherlock Holmes published the original photographs, along with an account of the verification process, in *The Strand Magazine*—the home publication of none other than Holmes himself. The title: "Fairies Photographed: An Epoch-Making Event." Two years later, he released a book, *The Coming of the Fairies*, which expanded on his initial investigation and included additional corroboration of the fairies' existence by the clairvoyant Mr. Geoffrey

Hodson. Conan Doyle had made up his mind, and he wasn't about to change it.

How had Conan Doyle failed the test of Holmesian thinking? What led such an obviously intelligent individual down a path to concluding that fairies existed simply because an expert had affirmed that the Cottingley photographs had not been faked?

Sir Arthur spent so much effort confirming the veracity of the photos that he never stopped to ask an obvious question: why, in all of the inquiries into whether the prints were genuine, did no one ask whether the fairies themselves might have been more easily manufactured? We can easily agree with the logic that it would seem improbable for a ten-year-old and a sixteen-year-old to fabricate photographs that could confound the experts, but what about fabricating a fairy? Take a look at the pictures on the preceding pages. It seems obvious in retrospect that they can't be real. Do those fairies look alive to you? Or do they more resemble paper cutouts, however artfully arranged? Why are they of such differing contrast? Why aren't their wings moving? Why did no one stay with the girls to see the fairies in person?

Conan Doyle could—and should—have dug deeper when it came to the young ladies in question. Had he done so, he would have discovered, for one, that young Elsie was a gifted artist—and one who, it just so happened, had been employed by a photography studio. He may have also discovered a certain book, published in 1915, whose pictures bore an uncanny resemblance to the fairies that appeared on the camera in the original prints.

Holmes surely wouldn't have been taken in so easily by the Cottingley photographs. Could the fairies have had human agents as well, agents who may have helped them get on camera, eased them into existence, so to speak? That would have been his first question. Something improbable is not yet impossible—but it requires a correspondingly large burden of proof. And that, it seems quite clear, was something Sir Arthur Conan Doyle did not quite provide. Why? As we will see, when we really want to believe something, we become far less skeptical and inquisitive, letting evidence pass muster with far less scrutiny than we would ever admit for a phenomenon we *didn't* want to believe. We don't, in other words, require as large or diligent a burden of proof. And for Conan Doyle, the existence of fairies was just such an instance.

When we make a decision, we decide within the context of knowledge that is available to us in the moment and not in retrospect. And within that context, it can be difficult indeed to balance the requisite open-mindedness with what passes for rationality *given the context of the times.* We, too, can be fooled into believing that fairies—or our version thereof—are real. All it takes is the right environment and the right motivation. Think of that before you leap to judge Conan Doyle's folly (something that, I hope, you will be less inclined to do before the chapter's end).

Prisoners of Our Knowledge and Motivation

Close your eyes and picture a tiger. It's lying on a patch of green grass, basking in the sun. It licks its paws. With a lazy yawn, it turns over onto its back. There's a rustle off to the side. It might just be the wind, but the tiger tenses up. In an instant, he is crouching on all fours, back arched, head drawn in between his shoulders.

Can you see it? What does it look like? What color is its fur? Does it have stripes? What color are those? What about the eyes? The face (are there whiskers)? The texture of the fur? Did you see its teeth when it opened its mouth?

If you're like most people, your tiger was a kind of orange, with dark black stripes lining its face and sides. Maybe you remembered to add the characteristic white spots to the face and underbelly, the tips of the paws and base of the neck. Maybe you didn't and your tiger was more mono-chrome than most. Maybe your tiger's eyes were black. Maybe they were blue. Both are certainly possible. Maybe you saw its incisors bared. Maybe you didn't.

But one detail is constant for nearly everyone: one thing your tiger was *not* is any predominant color other than that burnt orange-red hue that seems something between fire and molasses. It probably wasn't the rare white tiger, the albino-like creature whose white fur is caused by a double recessive gene that occurs so infrequently that experts estimate its natural incidence at only one out of approximately ten thousand tigers born in the wild. (Actually, they aren't albinos at all. The condition is called *leucism* and it results in a reduction of all skin pigments, not just melanin.) Nor is it likely to have been a black tiger, otherwise known as a melanistic tiger. That particular coloration—no stripes, no gradation, just pure, jet-black fur—is caused by a polymorphism that results in a non-agouti mutation (the agouti gene, essentially, determines whether a coat will be banded, the usual process of coloring each individual hair, or solid, non-agouti). Neither kind is common. Neither kind seems to be the typical *tiger* that the word brings to mind. And yet, all three are members of the exact same species, *panthera tigris*.

Now close your eyes and picture another animal: a mimic octopus.

It's perched on the ocean floor, near some reefs. The water is a misty blue. Nearby, a school of fish passes.

Stumped? Here's some help. This octopus is about two feet long, and has brown and white stripes or spots—except when it doesn't. You see, the mimic can copy over fifteen different sea animals. It can look like that jellyfish from "The Lion's Mane" that claimed so many victims right under the nose of a baffled Holmes. It can take the shape of a banded sea snake, a leaf-shaped sole, or something resembling a furry turkey with human legs. It can change color, size, and geometry all at a moment's notice. In other words, it's almost impossible to imagine it as any one thing. It is myriad animals at once, and none that you can pinpoint at any one instant.

Now I'm going to tell you one more thing. One of those animals mentioned in the preceding paragraphs doesn't actually exist. It may one day be real, but as of now it's the stuff of legend. Which one do you think it is? The orange tiger? The white one? The black one? The mimic octopus?

Here's the answer: the black tiger. While genetically it seems plausible—and what we know about the tiger's patterns of inheritance and genome confirms that it remains a theoretical possibility—a true melanistic tiger has never been seen. There have been allegations. There have been pseudo-melanistic examples (whose stripes are so thick and close as to almost give off the impression of melanism). There have been brown tigers with dark stripes. There have been black tigers that ended up being black leopards—the most common source of confusion. But there hasn't ever been a black tiger. Not one confirmed, verified case. Not ever.

And yet chances are you had little trouble believing in its existence. People have certainly wanted them to exist for centuries. The dark beasts figure in a Vietnamese legend; they've been the subject of numerous bounties; one was even presented as a gift to Napoleon from the king of Java (alas, it was a leopard). And they make sense. They fit in with the general pattern of animals that we expect to be real. And anyway, why ever not?

The mimic octopus, on the other hand, was indeed the stuff of legend until not too long ago. It was discovered only in 1998, by a group of fishermen off the coast of Indonesia. So strange was the report and so seemingly implausible that it took hours of footage to convince skeptical scientists that the creature was for real. After all, while mimicry is fairly common in the animal kingdom, never before had a single species been able to take on *multiple* guises—and never before had an octopus actually assumed the appearance of another animal.

The point is that it's easy to be fooled by seemingly scientific context into thinking something real when it's not. The more numbers we are given, the more details we see, the more we read big, scientific-seeming words like *melanism* instead of *plain black*, *agouti* and *non-agouti* instead of *banded* or *solid*, *mutation*, *polymorphism*, *allele*, *genetics*, piling them on word after word, the more likely we are to believe that the thing described is real. Conversely, it's all too easy to think that because something sounds implausible or out-there or discordant, because it has never before been seen and wasn't even suspected, it must be nonexistent.

Imagine for a moment that the Cottingley photographs had instead depicted the young girls with a never-before-seen variety of insect. What if, for instance, the picture had been of the girls handling this creature instead.

A miniature dragon, no less. (Actually, *draco sumatranus*, a gliding lizard native to Indonesia—but would anyone in England during Conan Doyle's time have been so wise?) Or this.

A creature of the deep, dark imagination, something out of a book of horrors, perhaps. But real? (Actually, the star-nosed mole, *condylura cristata*, is found in eastern Canada. Hardly common knowledge even in the pre-Internet days, let alone back in the Victorian era.)

Or indeed any number of animals that had seemed foreign and strange only decades earlier—and some that seem strange even today. Would they have been held to the same burden of proof—or would the lack of obvious fakery in the photograph have been enough?

What we believe about the world—and the burden of proof that we require to accept something as fact—is constantly shifting. These beliefs aren't quite the information that's in our brain attic, nor are they pure observation, but they are something that colors every step of the problem-solving process nevertheless. What we believe is possible or plausible shapes our basic assumptions in how we formulate and investigate questions. As we'll see, Conan Doyle was predisposed to believe in the possibility of fairies. He wanted them to be real. The predisposition in turn shaped his intuition about the Cottingley photographs, and that made all the difference in his failure to see through them, even though he acted with what he thought was great rigor in trying to establish their veracity.

An intuition colors how we interpret data. Certain things "seem" more plausible than others, and on the flip side, certain things just "don't make sense," no matter how much evidence there may be to support them. It's the confirmation bias (and many other biases at that: the illusion of validity and understanding, the law of small numbers, and anchoring and representativeness, all in one) all over again.

Psychologist Jonathan Haidt summarizes the dilemma in *The Righteous Mind*, when he writes, "We are terrible at seeking evidence that challenges our own beliefs, but other people do us this favor, just as we are good at finding errors in other people's beliefs." It's easy enough for most of us to spot the flaws in the fairies, because we have no emotional stake in their potential reality. But take something that touches us personally, where our very reputation might be on the line, and will it still be so simple?

It's easy to tell our minds stories about what is, and equally easy to tell them stories about what is not. It depends deeply on our motivation. Even still, we might think that fairies seem a far cry from a creature of the deep like the mimic octopus, no matter how hard it might be to fathom such a creature. After all, we know there are octopi. We know that new species of animals are discovered every day. We know some of them may seem a bit bizarre. Fairies, on the other hand, challenge every rational understanding we have of how the world works. And this is where context comes in.

A Recklessness of Mind?

Conan Doyle wasn't altogether reckless in authenticating the Cottingley photos. Yes, he did not gather the same exacting proof he would doubtless have demanded of his detective. (And it bears remembering that Sir Arthur was no slouch when it came to that type of thing. He was instrumental, you'll recall, in clearing the name of two falsely accused murder suspects, George Edalji and Oscar Slater.) But he did ask the best photography experts he knew. And he did try for replication—of a sort. And was it so difficult to believe that two girls of ten and sixteen would not be capable of the type of technical expertise that had been suggested as a means of falsifying the negatives?

It helps us to more clearly understand Conan Doyle's motivations if we try to see the photographs as he and his contemporaries would have seen them. Remember, this was before the age of digital cameras and Photoshopping and editing ad infinitum, when anyone can create just about anything that can be imagined—and do so in a much more convincing fashion than the Cottingley Fairies. Back then, photography was a relatively new art. It was labor intensive, time consuming, and technically challenging. It wasn't something that just anyone could do, let alone manipulate in a convincing fashion. When we look at the pictures today, we see them with different eyes than the eyes of 1920. We have different standards. We have grown up with different examples. There was a time when a photograph was considered high proof indeed, so difficult was it to take and to alter. It's nearly impossible to look back and realize how much has changed and how different the world once appeared.

Still, the Cottingley Fairies suffered from one major—and, it turned out for Conan Doyle's reputation, insurmountable—limitation. Fairies do not and cannot exist. It's just as that Kodak employee pointed out to Sir Arthur: the evidence *did not matter*, whatever it was. Fairies are creatures of the imagination and not of reality. End of story.

Our own view of what is and is not possible in reality affects how we perceive identical evidence. But that view shifts with time, and thus, evidence that might at one point seem meaningless can come to hold a great deal of meaning. Think of how many ideas seemed outlandish when first put forward, seemed so impossible that they couldn't be true: the earth being round; the earth going around the sun; the universe being made up almost entirely of something that we can't see, dark matter and energy. And don't forget that magical things *did* keep happening all around as Conan Doyle came of age: the invention of the X-ray (or the Röntgen ray, as it was called), the discovery of the germ, the microbe, radiation—all things that went from invisible and thus nonexistent to visible and apparent. Unseen things that no one had suspected were there were, in fact, very there indeed.

In that context, is it so crazy that Arthur Conan Doyle became a spiritualist? When he officially embraced Spiritualism in 1918, he was hardly alone in his belief—or knowledge, as he would have it. Spiritualism itself, while never mainstream, had prominent supporters on both sides of the ocean. William James, for one, felt that it was essential for the new discipline of psychology to test the possibilities of psychical research, writing: "Hardly, as yet, has the surface of the facts called 'psychic' begun to be scratched for scientific purposes. It is through following these facts, I am persuaded, that the greatest scientific conquests of the coming generation will be achieved." The psychic was the future, he thought, of the knowledge of the century. It was the way forward, not just for psychology, but for all of scientific conquest.

This from the man considered the father of modern psychology. Not to mention some of the other names who filled out the ranks of the psychical community. Physiologist and comparative anatomist William B. Carpenter, whose work included influential writings on comparative neurology; the renowned astronomer and mathematician Simon New-comb; naturalist Alfred Russel Wallace, who proposed the theory of evolution simultaneously with Charles Darwin; chemist and physicist William Crookes, discoverer of new elements and new methods for study-ing them; physicist Oliver Lodge, closely involved in the development of the wireless telegraph; psychologist Gustav Theodor Fechner, founder of one of the most precisely scientific areas of psychological research, psy-chophysics; physiologist Charles Richet, awarded the Nobel Prize for his work on anaphylaxis; and the list goes on.

And have we come that much further today? In the United States, as of 2004, 78 percent of people believed in angels. As for the spiritual realm as such, consider this. In 2011, Daryl Bem, one of the grand sires of mod-ern psychology—who made his name with a theory that contends that we perceive our own mental and emotional states much as we do others', by looking at physical signs—published a paper in the *Journal of Personality and Social Psychology*, one of the most respected and highly impactful publications in the discipline. The topic: proof of the existence of extra-sensory perception, or ESP. Human beings, he contends, can see the future.

In one study, for instance, Cornell University students saw two cur-tains on a screen. They had to say which curtain hid a picture. After they chose, the curtain was opened, and the researcher would show them the picture's location.

What's the point, you might (reasonably enough) wonder, to show a location *after* you've already made your choice? Bem argues that if we are able to see even a tiny bit into the future, we will be able to retroactively use that information to make better-than-average guesses in the present.

It gets even better. There were two types of photographs: neutral ones, and ones showing erotic scenes. In Bem's estimation, there was a chance that we'd be better at seeing the future if it was worth seeing (wink, wink, nudge, nudge). If he was correct, we'd be better than the fifty-fifty pre-dicted by chance at guessing the image. Lo and behold, rates for the erotic images hovered around 53 percent. ESP is real. Everyone, rejoice. Or, in the more measured words of psychologist Jonathan Schooler (one of the reviewers of the article), "I truly believe that this kind of finding from a well-respected, careful researcher deserves public airing." It's harder than we thought to leave the land of fairies and Spiritualism behind. It's all the more difficult to do when it deals with something we *want* to believe.

Bem's work has launched the exact same cries of "crisis of the discipline" that arose with William James's public embrace of Spiritualism over one hundred years ago. In fact, it is called out as such in the very same issue that carries the study—a rare instance of article and rebuttal appearing simultaneously. Might JPSP have seen the future and tried to stay a step ahead of the controversial decision to publish at all?

Not much has changed. Except now, instead of psychical research and Spiritualism it's called psi, parapsychology, and ESP. (On the flip side, how many people refuse to believe Stanley Milgram's results on obedience, which showed that the vast majority of people will deliver lethal levels of shock when ordered to do so, with full knowledge of what they are doing, even when confronted with them?) Our instincts are tough to beat, whichever way they go. It takes a mindful effort of will.

Our intuition is shaped by context, and that context is deeply informed by the world we live in. It can thus serve as a blinder—or blind spot—of sorts, much as it did for Conan Doyle and his fairies. With mindfulness, however, we can strive to find a balance between fact-checking our intuitions and remaining open-minded. We can then make our best judgments, with the information we have and no more, but with, as well, the understanding that time may change the shape and color of that information.

Can we really blame, then, Arthur Conan Doyle's devotion to his fairy stories? Against the backdrop of Victorian England, where fairies populated the pages of nigh every children's book (not least of all *Peter Pan*, by Sir Arthur's own good friend J. M. Barrie), where even the physicists and psychologists, the chemists and the astronomers were willing to grant that there might be something to it, was he so far off? After all, he was only human, just like us.

We will never know it all. The most we can do is remember Holmes's precepts and apply them faithfully. And to remember that open-mindedness is one of them—hence the maxim (or axiom, as he calls it on this particular occasion in "The Adventure of the Bruce-Partington Plans"), "When all other contingencies fail, whatever remains, however improbable, must be the truth."

But how do we do this in practice? How do we go beyond theoretically understanding this need for balance and open-mindedness and applying it practically, in the moment, in situations where we might not have as much time to contemplate our judgments as we do in the leisure of our reading?

It all goes back to the very beginning: the habitual mindset that we cultivate, the structure that we try to maintain for our brain attic no matter what.

The Mindset of a Hunter

One of the images of Sherlock Holmes that recurs most often in the stories is that of Holmes the hunter, the ever-ready predator looking to capture his next prey even when he appears to be lounging calmly in the shade, the vigilant marksman alert to the slightest activity even as he balances his rifle across his knees during a midafternoon break.

Consider Watson's description of his companion in "The Adventure of the Devil's Foot."

> One realized the red-hot energy which underlay Holmes's phlegmatic exterior when one saw the sudden change which came over him from the moment that he entered the fatal apartment. In an instant he was tense and alert, his eyes shining, his face set, his limbs quivering with eager activity . . . for all the world like a dashing foxhound drawing a cover.

It's the perfect image, really. No energy wasted needlessly, but an ever-alert, habitual state of attention that makes you ready to act at a moment's notice, be it as a hunter who has glimpsed a lion, a lion who has glimpsed a gazelle, or a foxhound who has sensed the fox near and whose body has become newly alerted to the pursuit. In the symbol of the hunter, all of the qualities of thought that Sherlock Holmes epitomizes merge together into a single, elegant shape. And in cultivating that mind-set, in all of its precepts, we come one step closer to being able to do in practice what we understand in theory. The mind of a hunter encapsu-lates the elements of Holmesian thought that might otherwise get away from us, and learning to use that mindset regularly can remind us of principles that we might otherwise let slide.

Ever-Ready Attention

Being a hunter doesn't mean always hunting. It means always being ready to go on alert, when the circumstances warrant it, but not squandering your energy needlessly when they don't. Being attuned to the signs that need attending to, but knowing which ones to ignore. As any good hunter knows, you need to gather up your resources for the moments that matter.

Holmes's lethargy—that "phlegmatic exterior" that in others might signal melancholy or depression or pure laziness—is calculated. There is nothing lethargic about it. In those deceptive moments of inaction, his energy is pent up in his mind attic, circulating around, peering into the corners, gathering its strength in order to snap into focus the moment it is called on to do so. At times, the detective even refuses to eat because he doesn't want to draw blood from his thoughts. "The faculties become re-fined when you starve them," Holmes tells Watson in "The Adventure of the Mazarin Stone," when Watson urges him to consume at least some food. "Why, surely, as a doctor, my dear Watson, you must admit that what your digestion gains in the way of blood supply is so much lost to the brain. I am a brain, Watson. The rest of me is a mere appendix. Therefore, it is the brain I must consider."

We can never forget that our attention—and our cognitive abilities more broadly—are part of a finite pool that will dry out if not managed properly and replenished regularly. And so, we must employ our atten-tional resources mindfully—and selectively. Be ready to pounce when that tiger does make an appearance, to tense up when the scent of the fox car-ries on the breeze, the same breeze that to a less attentive nose than yours signifies nothing but spring and fresh flowers. Know when to engage, when to withdraw—and when something is beside the point entirely.

Environmental Appropriateness

A hunter knows what game he is hunting, and he modifies his approach accordingly. After all, you'd hardly hunt a fox as you would a tiger, approach the shooting of a partridge as you would the stalking of a deer. Unless you're content with hunting the same type of prey over and over, you must learn to be appropriate to the circumstances, to modify your weapon, your approach, your very demeanor according to the dictates of the specific situation.

Just as a hunter's endgame is always the same—kill the prey—Holmes's goal is always to obtain information that will lead him to the suspect. And yet, consider how Holmes's approach differs depending on the person he is dealing with, the specific "prey" at hand. He reads the person, and he proceeds accordingly.

In "The Adventure of the Blue Carbuncle," Watson marvels at Holmes's ability to get information that, only moments earlier, was not forthcoming. Holmes explains how he was able to do it: "When you see a man with whiskers of that cut and the 'Pink 'un' protruding out of his pocket, you can always draw him by a bet," said he. "I daresay that if I had put £100 down in front of him, that man would not have given me such complete information as was drawn from him by the idea that he was doing me on a wager."

Contrast this tactic with that employed in *The Sign of Four*, when Holmes sets out to learn the particulars of the steam launch *Aurora*. "The main thing with people of that sort," he tells Watson, "is never to let them think that their information can be of the slightest importance to you. If you do they will instantly shut up like an oyster. If you listen to them under protest, as it were, you are very likely to get what you want."

You don't bribe someone who thinks himself above it. But you do approach him with a bet if you see the signs of betting about his person. You don't hang on to every word with someone who doesn't want to be giving information to just anybody. But you do let them prattle along and pretend to indulge them if you see any tendency to gossip. Every person is different, every situation requires an approach of its own. It's the reckless hunter indeed who goes to hunt the tiger with the same gun he reserves for the pheasant shoot. There is no such thing as one size fits all. Once you have the tools, once you've mastered them, you can wield them with greater authority and not use a hammer where a gentle tap would do. There's a time for straightforward methods, and a time for more unorthodox ones. The hunter knows which is which and when to use them.

Adaptability

A hunter will adapt when his circumstances change in an unpredictable fashion. What if you should be out hunting ducks and just so happen to spot a deer in the nearby thicket? Some may say, *No thanks*, but many would adapt to the challenge, using the opportunity to get at a more valuable, so to speak, prey.

Consider "The Adventure of the Abbey Grange," when Holmes decides at the last moment not to give up the suspect to Scotland Yard. "No I couldn't do it, Watson," he says to the doctor.

> "Once that warrant was made out, nothing on earth would save him. Once or twice in my career I feel that I have done more real harm by my discovery of the criminal than ever he had done by his crime. I have learned caution now, and I had rather play tricks with the law of England than with my own conscience. Let us know a little more before we act."

You don't mindlessly follow the same preplanned set of actions that you had determined early on. Circumstances change, and with them so does the approach. You have to think before you leap to act, or to judge someone, as the case may be. Everyone makes mistakes, but some may not be mistakes as such, when taken in context of the time and the situation. (After all, we wouldn't make a choice if we didn't think it the right one at the time.) And if you do decide to keep to the same path, despite the changes, at least you will choose the so-called nonoptimal route mindfully, and with full knowledge of why you're doing it. And you will learn to always "know a little more" before you act. As William James puts it, "We all, scientists and non-scientists, live on some inclined plane of credulity. The plane tips one way in one man, another way in another; and may he whose plane tips in no way be the first to cast a stone!"

Acknowledging Limitations

The hunter knows his weak spots. If he has a blind side, he asks someone to cover it; or he makes sure it is not exposed, if no one is available. If he tends to overshoot, he knows that, too. Whatever the handicap, he must take it into account if he is to emerge successful from the hunt.

In "The Disappearance of Lady Frances Carfax," Holmes realizes where the eponymous lady has disappeared to only when it is almost too late to save her. "Should you care to add the case to your annals, my dear Watson," he says, once they return home, having beaten the clock by mere minutes, "it can only be as an example of that temporary eclipse to which even the best-balanced mind may be exposed. Such slips are common to all mortals, and the greatest is he who can recognize and repair them. To this modified credit I may, perhaps, make some claim."

The hunter must err before he realizes where his weakness may lie. The difference between the successful hunter and the unsuccessful one isn't a lack of error. It is the recognition of error, and the ability to learn from it and to prevent its occurrence in the future. We need to recognize our limitations in order to overcome them, to know that we are fallible, and to recognize the fallibility that we see so easily in others in our own thoughts and actions. If we don't, we'll be condemned to always believe in fairies—or to never believe in them, even should signs point to the need for a more open-minded consideration.

Cultivating Quiet

A hunter knows when to quiet his mind. If he allows himself to always take in everything that is there for the taking, his senses will become overwhelmed. They will lose their sharpness. They will lose their ability to focus on the important signs and to filter out the less so. For that kind of vigilance, moments of solitude are essential.

Watson makes the point succinctly in *The Hound of the Baskervilles*, when Holmes asks to be left alone. His friend doesn't complain. "I knew that seclusion and solitude were very necessary for my friend in those hours of intense mental concentration during which he weighed every particle of evidence, constructed alternative theories, balanced one against the other, and made up his mind as to which points were essential and which immaterial," he writes.

The world is a distracting place. It will never quiet down for you, nor will it leave you alone of its own accord. The hunter must seek out his own seclusion and solitude, his own quietness of mind, his own space in which to think through his tactics, his approaches, his past actions, and his future plans. Without that occasional silence, there can be little hope of a successful hunt.

Constant Vigilance

And most of all, a hunter never lets down his guard, not even when he thinks that no tiger in its right mind could possibly be out and about in the heat of the afternoon sun. Who knows, it might just be the day that the first-ever black tiger is spotted, and that tiger may have different hunting habits than you are used to (isn't its camouflage different? wouldn't it make sense that it would approach in an altogether different manner?). As Holmes warns over and over, it is the least remarkable crime that is often the most difficult. Nothing breeds complacency like routine and the semblance of normality. Nothing kills vigilance so much as the commonplace. Nothing kills the successful hunter like a complacency bred of that very success, the polar opposite of what enabled that success to begin with.

Don't be the hunter who missed his prey because he thought he'd gotten it all down so well that he succumbed to mindless routine and action. Remain ever mindful of how you apply the rules. Never stop thinking.

It's like the moment in *The Valley of Fear* when Watson says, "I am inclined to think—" and Holmes cuts him off in style: "I should do so."

Could there be a more appropriate image to that awareness of mind that is the pinnacle of the Holmesian approach to thought? A brain, first and foremost, and in it, the awareness of a hunter. The hunter who is never just inclined to think, but who does so, always. For that mindfulness doesn't begin or end with the start of each hunt, the beginning of each new venture or thought process. It is a constant state, a well-rehearsed presence of mind even as he settles down for the night and stretches his legs in front of the fire.

Learning to think like a hunter will go a long way toward making sure that we don't blind ourselves to the obvious inconsistencies of fairy land when they stare us in the face. We shouldn't rule them out, but we should be wary—and know that even if we really want to be the ones to discover the first real proof of their existence, that proof may still be in the future, or nowhere at all; in either case, the evidence should be treated just as severely. And we should apply that same attitude to others and their beliefs.

The way you see yourself matters. View yourself as a hunter in your own life, and you may find yourself becoming more able to hunt properly, in a matter of speaking. Whether you choose to consider the possibility of fairies' existence or not, you—the hunter you—will have done it thinkingly. You won't have been unprepared.

In 1983, the tale of the Cottingley Fairies came to as near an end as it ever would. More than sixty years after the photographs first surfaced, seventy-six-year-old Frances Griffiths made a confession: the photographs were fake. Or at least four of them were. The fairies had been her older cousin's illustrations, secured by hat pins to the scenery. And the evidence of a belly button that Conan Doyle had thought he'd seen on the goblin in the original print was actually nothing more than that—a hat pin. The final photograph, however, was genuine. Or so said Frances.

Two weeks later, Elsie Hill (née Wright) herself came forward. It's true, she said, after having held her silence since the original incident.

She had drawn the fairies in sepia on Windsor and Bristol board, coloring them in with watercolors while her parents were out of the house. She had then fastened them to the ground with hat pins. The figures themselves had apparently been traced from the 1915 *Princess Mary Gift Book*. And that last picture, that Frances had maintained was real? Frances wasn't even there, Elsie told *The Times*. "I am very proud of that one—it was all done with my own contraption and I had to wait for the weather to be right to take it," she said. "I won't reveal the secret of that one until the very last page of my book."

Alas, the book was never written. Frances Griffiths died in 1986 and Elsie, two years later. To this day, there are those who maintain that the fifth photograph was genuine. The Cottingley Fairies just refuse to die.

But maybe, just maybe, Conan Doyle the hunter would have escaped the same fate. Had he taken himself (and the girls) just a bit more critically, pried just a bit harder, perhaps he could have learned from his mistakes, as did his creation when it came to his own vices. Arthur Conan Doyle may have been a Spiritualist, but his spirituality failed to take the one page of Sherlock Holmes that was nonnegotiable for the taking: mindfulness.

W. H. Auden writes of Holmes,

His attitude towards people and his technique of observation and deduction are those of the chemist or physicist. If he chooses human beings rather than inanimate matter as his material, it is because investigating the inanimate is unheroically easy since it cannot tell lies, which human beings can and do, so that in dealing with them, observation must be twice as sharp and logic twice as rigorous.

Sir Arthur Conan Doyle valued few things as highly as he did heroism. And yet he failed to realize that the animals he was hunting were just as human as those that he created. He was not twice as sharp, twice as logical, twice as rigorous. But perhaps he could have been, with a little help from the mindset that he himself created for his own detective, someone who would surely have never forgotten that human beings can and do tell lies, that everyone can be mistaken and everyone is fallible, ourselves included.

Conan Doyle could not know where science was headed. He did the best he could, and did so within the parameters that he had set for himself, and which, I might add, remain to this day. For, unlike William James's confident prediction, our knowledge about the unseen forces that guide our lives, while light-years further than Sir Arthur could ever imagine when it comes to explaining natural phenomena, is still stuck circa 1900 when it comes to explaining psychical ones.

But the point is greater than either Sherlock Holmes or Arthur Conan Doyle—or, for that matter, Daryl Bem or William James. We are all limited by our knowledge and context. And we'd do well to remember it. Just because we can't fathom something doesn't make it not so. And just because we screw up for lack of knowledge doesn't mean we've done so irredeemably—or that we can't keep learning. When it comes to the mind, we can all be hunters.

SHERLOCK HOLMES FURTHER READING

"And yet the motives of women are so inscrutable" from *The Return of Sherlock Holmes,* "The Adventure of the Second Stain," p. 1189.

"If the devil did decide to have a hand in the affairs of men—." "I knew that seclusion and solitude were very necessary for my friend . . ." from *The Hound of the Baskervilles,* chapter 3: The Problem, p. 22.

"One realized the red-hot energy that underlay Holmes's phlegmatic exterior." from *His Last Bow,* "The Adventure of the Devil's Foot," p. 1392.

"When you see a man with whiskers of that cut and the 'pink 'un' protruding out of his pocket, you can always draw him by a bet." from *The Adventures of Sherlock Holmes,* "The Adventure of the Blue Carbuncle," p. 158.

"Once that warrant was made out, nothing on earth could save him." from *The Return of Sherlock Holmes,* "The Adventure of the Abbey Grange," p. 1158.

"Should you care to add the case to your annals, my dear Watson, it can only be as an example of that temporary eclipse to which even the best-balanced mind may be exposed." from *His Last Bow,* "The Disappearance of Lady Frances Carfax," p. 342.

"I am inclined to think—." from *The Valley of Fear,* Part One, chapter 1: The Warning, p. 5.

Postlude

Walter Mischel was nine years old when he started kindergarten. It wasn't that his parents had been negligent in his schooling. It was just that the boy couldn't speak English. It was 1940 and the Mischels had just arrived in Brooklyn. They'd been one of the few Jewish families lucky enough to escape Vienna in the wake of the Nazi takeover in the spring of 1938. The reason had as much to do with luck as with foresight: they had discovered a certificate of U.S. citizenship from a long-since-dead maternal grandfather. Apparently, he had obtained it while working in New York City around 1900, before returning once more to Europe.

But ask Dr. Mischel to recall his earliest memories, and chances are that the first thing he will speak of is not how the Hitler Youths stepped on his new shoes on the sidewalks of Vienna. Nor will it be of how his father and other Jewish men were dragged from their apartments and forced to march in the streets in their pajamas while holding branches in their hands, in a makeshift "parade" staged by the Nazis in parody of the Jewish tradition of welcoming spring. (His father had polio and couldn't walk without his cane. And so, the young Mischel had to watch as he jerked from side to side in the procession.) Nor will it be of the trip from Vienna, the time spent in London in an uncle's spare room, the journey to the United States at the outbreak of war.

Instead, it will be of the earliest days in that kindergarten classroom, when little Walter, speaking hardly a word of English, was given an IQ test. It should hardly come as a surprise that he did not fare well. He was in an alien culture and taking a test in an alien language. And yet his teacher *was* surprised. Or so she told him. She also told him how disappointed she was. Weren't foreigners supposed to be smart? She'd expected more from him.

Carol Dweck was on the opposite side of the story. When she was in sixth grade—also, incidentally, in Brooklyn—she, too, was given an intelligence test, along with the rest of her class. The teacher then proceeded to do something that today would raise many eyebrows but back then was hardly uncommon: she arranged the students in order of score. The "smart" students were seated closest to the teachers. And the less fortunate, farther and farther away. The order was immutable, and those students who had fared less than well weren't even allowed to perform such basic classroom duties as washing the blackboard or carrying the flag to the school assembly. They were to be reminded constantly that their IQ was simply not up to par.

Dweck herself was one of the lucky ones. Her seat: number one. She had scored highest of all her classmates. And yet, something wasn't quite right. She knew that all it would take was another test to make her less smart. And could it be that it was so simple as all that—a score, and then your intelligence was marked for good?

Years later, Walter Mischel and Carol Dweck both found themselves on the faculty of Columbia University. (As of this writing, Mischel is still there and Dweck has moved to Stanford.) Both had become key players in social and personality psychology research (though Mischel the sixteen-years-senior one), and both credit that early test to their subsequent career trajectories, their desire to conduct research into such supposedly fixed things as personality traits and intelligence, things that could be measured with a simple test and, in that measurement, determine your future.

It was easy enough to see how Dweck had gotten to that pinnacle of academic achievement. She was, after all, the smartest. But what of Mischel? How could someone whose IQ would have placed him squarely in the back of Dweck's classroom have gone on to become one of the leading figures in psychology of the twentieth century, he of the famous marshmallow studies of self-control and of an entirely new approach to looking at personality and its measurement? Something wasn't quite right, and the fault certainly wasn't with Mischel's intelligence or his stratospheric career trajectory.

Sherlock Holmes is a hunter. He knows that there is nothing too difficult for his mastery—in fact, the more difficult something is, the better. And in that attitude may lie a large portion of his success, and a large part of Watson's failure to follow in his footsteps. Remember that scene from "The Adventure of the Priory School," where Watson all but gives up hope at figuring out what happened to the missing student and teacher?

"I am at my wit's end," he tells Holmes.

But Holmes will have none of it. "Tut, tut, we have solved worse problems."

Or, consider Holmes's response to Watson when the latter declares a cipher "beyond human power to penetrate."

Holmes answers, "Perhaps there are points that have escaped your Machiavellian intellect." But Watson's attitude is surely not helping. "Let us continue the problem in the light of pure reason," he directs him, and goes on, naturally, to decipher the note.

In a way, Watson has beaten himself in both cases before he has even started. By declaring himself at his wit's end, by labeling something as beyond human power, he has closed his mind to the possibility of success. And that mindset, as it turns out, is precisely what matters most—and it's a thing far more intangible and unmeasurable than a number on a test.

For many years, Carol Dweck has been researching exactly what it is that separates Holmes's "tut, tut" from Watson's "wit's end," Walter Mischel's success from his supposed IQ. Her research has been guided by two main assumptions: IQ cannot be the only way to measure intelligence, and there might be more to that very concept of intelligence than meets the eye.

According to Dweck, there are two main theories of intelligence: incremental and entity. If you are an incremental theorist, you believe that intelligence is fluid. If you work harder, learn more, apply yourself better, you will become smarter. In other words, you dismiss the notion that something might possibly be beyond human power to penetrate. You think that Walter Mischel's original IQ score is not only something that should not be a cause for disappointment but that it has little bearing on his actual ability and later performance.

If, on the other hand, you are an entity theorist, you believe that intelligence is fixed. Try as you might, you will remain as smart (or not) as you were before. It's just your original luck. This was the position of Dweck's sixth-grade teacher—and of Mischel's kindergarten one. It means that once in the back, you're stuck in the back. And there's nothing you can do about it. Sorry, buddy, luck of the draw.

In the course of her research, Dweck has repeatedly found an interesting thing: how someone performs, especially in reacting to failure, largely depends on which of the two beliefs he espouses. An incremental theorist sees failure as a learning opportunity; an entity theorist, as a frustrating personal shortcoming that cannot be remedied. As a result, while the former may take something away from the experience to apply to future situations, the latter is more likely to write it off entirely. So basically, how we think of the world and of ourselves can actually change how we learn and what we know.

In a recent study, a group of psychologists decided to see if this differential reaction is simply behavioral, or if it actually goes deeper, to the level of brain performance. The researchers measured response-locked event-related potentials (ERPs)—electric neural signals that result from either an internal or external event—in the brains of college students as they took part in a simple flanker task. The students were shown a string of five letters and asked to quickly identify the middle letter. The letters could be congruent—for instance, MMMMM—or they might be incongruent—for example, MMNMM.

While performance accuracy was generally high, around 91 percent, the specific task parameters were hard enough that everyone made some mistakes. But where individuals differed was in how both they—and, crucially, their brains—responded to the mistakes. Those who had an incremental mindset (i.e., believed that intelligence was fluid) performed better following error trials than those who had an entity mindset (i.e., believed intelligence was fixed). Moreover, as that incremental mindset increased, positivity ERPs on error trials as opposed to correct trials increased as well. And the larger the error positivity amplitude on error trials, the more accurate the post-error performance.

So what exactly does that mean? From the data, it seems that a growth mindset, whereby you believe that intelligence can improve, lends itself to a more adaptive response to mistakes—not just behaviorally but neurally. The more someone believes in improvement, the larger the amplitude of a brain signal that reflects a conscious allocation of attention to errors. And the larger that neural signal, the better the subsequent performance. That mediation suggests that individuals with an incremental theory of intelligence may actually have better self-monitoring and control systems on a very basic neural level: their brains are better at monitoring their own, self-generated errors and at adjusting their behavior accordingly. It's a story of improved online error awareness—of noticing mistakes as they happen, and correcting for them immediately.

The way our brains act is infinitely sensitive to the way we, their owners, think. And it's not just about learning. Even something as theoretical as belief in free will can change how our brains respond (if we don't believe in it, our brains actually become more lethargic in their preparation). From broad theories to specific mechanisms, we have an uncanny ability to influence how our minds work, and how we perform, act, and interact as a result. If we think of ourselves as able to learn, learn we will. And if we think we are doomed to fail, we doom ourselves to do precisely that, not just behaviorally but at the most fundamental level of the neuron.

But mindset isn't predetermined, just as intelligence isn't a monolithic thing that is preset from birth. We can learn, we can improve, we can change our habitual approach to the world. Take the example of stereotype threat, an instance where others' perception of us—or what we think that perception is—influences how we in turn act, and does so on the same subconscious level as all primes. Being a token member of a group (for example, a single woman among men) can increase self-consciousness and negatively impact performance. Having to write down your ethnicity or gender before taking a test has a negative impact on math scores for females and overall scores for minorities. (On the GREs, for instance, having race made salient lowers black students' performance.) Asian women perform better on a math test when their Asian identity is made salient, and worse when their female identity is. White men perform worse on athletic tasks when they think performance is based on natural ability, and black men when they are told it is based on athletic intelligence. It's called stereotype threat.

But a simple intervention can help. Women who are given examples of females successful in scientific and technical fields don't experience the negative performance effects on math tests. College students exposed to Dweck's theories of intelligence—specifically, the incremental theory—have higher grades and identify more with the academic process at the end of the semester. In one study, minority students who wrote about the personal significance of a self-defining value (such as family relationships or musical interests) three to five times during the school year had a GPA that was 0.24 grade points higher over the course of two years than those who wrote about neutral topics—and low-achieving African Americans showed improvements of 0.41 points on average. Moreover, the rate of remediation dropped from 18 percent to 5 percent.

What is the mindset you typically have when it comes to yourself? If you don't realize you have it, you can't do anything to combat the influences that come with it when they are working against you, as happens with negative stereotypes that hinder performance, and you can't tap into the benefits when they are working for you (as can happen if you activate positively associated stereotypes). What we believe is, in large part, how we are.

It is an entity world that Watson sees when he declares himself beaten–black and white, you know it or you don't, and if you come up against something that seems too difficult, well, you may as well not even try lest you embarrass yourself in the process. As for Holmes, everything is incremental. You can't know if you haven't tried. And each challenge is an opportunity to learn something new, to expand your mind, to improve your abilities and add more tools to your attic for future use. Where Watson's attic is static, Holmes's is dynamic.

Our brains never stop growing new connections and pruning unused ones. And they never stop growing stronger in those areas where we strengthen them, like that muscle we encountered in the early pages of the book, that keeps strengthening with use (but atrophies with disuse), that can be trained to perform feats of strength we'd never before thought possible.

How can you doubt the brain's transformational ability when it comes to something like thinking when it is capable of producing talent of all guises in people who had never before thought they had it in them? Take the case of the artist Ofey. When Ofey first started to paint, he was a middle-aged physicist who hadn't drawn a day in his life. He wasn't sure he'd ever learn how. But learn he did, going on to have his own one-man show and to sell his art to collectors all over the world.

Ofey, of course, is not your typical case. He wasn't just any physicist. He happens to have been the Nobel Prize–winning Richard Feynman, a man of uncommon genius in nearly all of his pursuits. Feynman had created Ofey as a pseudonym to ensure that his art was valued on its own terms and not on those of his laurels elsewhere. And yet there are multiple other cases. While Feynman may be unique in his contributions to physics, he certainly is not in representing the brain's ability to change—and to change in profound ways—late in life.

Anna Mary Robertson Moses—better known as Grandma Moses—did not begin to paint until she was seventy-five. She went on to be compared to Pieter Bruegel in her artistic talent. In 2006, her painting *Sugaring Off* sold for $1.2 million.

Václav Havel was a playwright and writer—until he became the center of the Czech opposition movement and then the first post-Communist president of Czechoslovakia at the age of fifty-three.

Richard Adams did not publish *Watership Down* until he was fifty-two. He'd never even thought of himself as a writer. The book that was to sell over fifty million copies (and counting) was born out of a story that he told to his daughters.

Harlan David Sanders—better known as Colonel Sanders—didn't start his Kentucky Fried Chicken company until the age of sixty-five, but he went on to become one of the most successful businessmen of his generation.

The Swedish shooter Oscar Swahn competed in his first Olympic games in 1908, when he was sixty years old. He won two gold and one bronze medals, and when he turned seventy-two, he became the oldest Olympian ever and the oldest medalist in history after his bronze-winning performance at the 1920 games. The list is long, the examples varied, the accomplishments all over the map.

And yes, there are the Holmeses who have the gift of clear thought from early on, who don't have to change or strike out in a new direction after years of bad habits. But never forget that even Holmes had to train himself, that even he was not born thinking like Sherlock Holmes. Nothing just happens out of the blue. We have to work for it. But with proper attention, it happens. It is a remarkable thing, the human brain.

As it turns out, Holmes's insights can apply to most anything. It's all about the attitude, the mindset, the habits of thinking, the enduring approach to the world that you develop. The specific application itself is far less important.

If you get only one thing out of this book, it should be this: the most powerful mind is the quiet mind. It is the mind that is present, reflective, mindful of its thoughts and its state. It doesn't often multitask, and when it does, it does so with a purpose.

The message may be getting across. A recent *New York Times* piece spoke of the new practice of squatting while texting: remaining in parked cars in order to engage in texting, emailing, Twittering, or whatever it is you do instead of driving off to vacate parking spaces. The practice may provoke parking rage for people looking for spots, but it also shows an increased awareness that doing anything while driving may not be the best idea. "It's time to kill multitasking" rang a headline at the popular blog *The 99%*.

We can take the loudness of our world as a limiting factor, an excuse as to why we cannot have the same presence of mind that Sherlock Holmes did—after all, he wasn't constantly bombarded by media, by technology, by the ever more frantic pace of modern life. He had it so much easier. Or, we can take it as a challenge to do Holmes one better. To show that it doesn't really matter—we can still be just as mindful as he ever was, and then some, if only we make the effort. And the greater the effort, we might say, the greater the gain and the more stable the shift in habits from the mindless toward the mindful.

We can even embrace technology as an unexpected boon that Holmes would have been all too happy to have. Consider this: a recent study demonstrated that when people are primed to think about computers, or when they expect to have access to information in the future, they are far less able to recall the information. However—and this is key—they are far better able to remember where (and how) to find the information at a later point.

In the digital age, our mind attics are no longer subject to the same constraints as were Holmes's and Watson's. We've in effect expanded our storage space with a virtual ability that would have been unimaginable in Conan Doyle's day. And that addition presents an intriguing opportunity. We can store "clutter" that might be useful in the future and know exactly how to access it should the need arise. If we're not sure whether something deserves a prime spot in the attic, we need not throw it out. All we need to do is remember that we've stored it for possible future use. But with the opportunity comes the need for caution. We might be tempted to store outside our mind attics that which should rightly be *in* our mind attics, and the curatorial process (what to keep, what to toss) becomes increasingly difficult.

Holmes had his filing system. We have Google. We have Wikipedia. We have books and articles and stories from centuries ago to the present day, all neatly available for our consumption. We have our own digital files.

But we can't expect to consult everything for every choice that we make. Nor can we expect to remember everything that we are exposed to—and the thing is, we shouldn't want to. We need to learn instead the art of curating our attics better than ever. If we do that, our limits have indeed been expanded in unprecedented ways. But if we allow ourselves to get bogged down in the morass of information flow, if we store the irrelevant instead of those items that would be best suited to the limited storage space that we always carry with us, in our heads, the digital age can be detrimental.

Our world is changing. We have more resources than Holmes could have ever imagined. The confines of our mind attic have shifted. They have expanded. They have increased the sphere of the possible. We should strive to be cognizant of that change, and to take advantage of the shift instead of letting it take advantage of us. It all comes back to that very basic notion of attention, of presence, of mindfulness, of the mindset and the motivation that accompany us throughout out lives.

We will never be perfect. But we can approach our imperfections mindfully, and in so doing let them make us into more capable thinkers in the long term.

"Strange how the brain controls the brain!" Holmes exclaims in "The Adventure of the Dying Detective." And it always will. But just maybe we can get better at understanding the process and lending it our input.

ACKNOWLEDGMENTS

So many extraordinary people have helped to make this book possible that it would take another chapter—at the very least; I'm not always known for my conciseness—to thank them all properly. I am incredibly grateful to everyone who has been there to guide and support me throughout it all: to my family and wonderful friends, I love you all and wouldn't have even gotten started, let alone finished, with this book without you; and to all of the scientists, researchers, scholars, and Sherlock Holmes aficionados who have helped guide me along the way, a huge thank you for your tireless assistance and endless expertise.

I'd like to thank especially Steven Pinker, the most wonderful mentor and friend I could ever imagine, who has been selfless in sharing his time and wisdom with me for close to ten years (as if he had nothing better to do). His books were the reason I first decided to study psychology—and his support is the reason I am still here. Richard Panek, who helped shepherd the project from its inception through to its final stages, and whose advice and tireless assistance were essential to getting it off the ground (and keeping it there). Katherine Vaz, who has believed in my writing from the very beginning and has remained for many years a constant source of encouragement and inspiration. And Leslie Klinger, whose early interest in my work on Mr. Holmes and unparalleled expertise on the world of 221B Baker Street were essential to the success of the journey.

My amazing agent, Seth Fishman, deserves constant praise; I'm lucky to have him on my side. Thank you to the rest of the team at the Gernert Company—and a special thanks to Rebecca Gardner and Will Roberts. My wonderful editors, Kevin Doughten and Wendy Wolf, have taken the manuscript from nonexistent to ready-for-the-world in under a year—something I never thought possible. I'm grateful as well to the rest of the team at Viking/Penguin, especially Yen Cheong, Patricia Nicolescu, Veronica Windholz, and Brittney Ross. Thank you to Nick Davies for his insightful edits and to everyone at Canongate for their belief in the project.

This book began as a series of articles in *Big Think* and *Scientific American*. A huge thank you to Peter Hopkins, Victoria Brown, and everyone at *Big Think* and to Bora Zivkovic and everyone at *Scientific American* for giving me the space and freedom to explore these ideas as I wanted to.

Far more people than I could list have been generous with their time, support, and encouragement throughout this process, but there are a few in particular I would like to thank here: Walter Mischel, Elizabeth Greenspan, Lyndsay Faye, and all of the lovely ladies of ASH, everyone at the Columbia University Department of Psychology, Charlie Rose, Harvey Mansfield, Jenny 8. Lee, Sandra Upson, Meg Wolitzer, Meredith Kaffel, Allison Lorentzen, Amelia Lester, Leslie Jamison, Shawn Otto, Scott Hueler, Michael Dirda, Michael Sims, Shara Zaval, and Joanna Levine.

Last of all, I'd like to thank my husband, Geoff, without whom none of this would be possible. I love you and am incredibly lucky to have you in my life.

FURTHER READING

The further reading sections at the end of each chapter reference page numbers from the following editions:

Conan Doyle, Arthur. (2009). *The Adventures of Sherlock Holmes*. Penguin Books: New York.

Conan Doyle, Arthur. (2001). *The Hound of the Baskervilles*. Penguin Classics: London.

Conan Doyle, Arthur. (2011). *The Memoirs of Sherlock Holmes*. Penguin Books: New York.

Conan Doyle, Arthur. (2001). *The Sign of Four*. Penguin Classics: London.

Conan Doyle, Arthur. (2001). *A Study in Scarlet*. Penguin Classics: London.

Conan Doyle, Arthur. (2001). *The Valley of Fear and Selected Cases*. Penguin Classics: London.

Conan Doyle, Arthur. (2005). *The New Annotated Sherlock Holmes*. Ed. Leslie S. Klinger. Norton: New York. Vol. II.

In addition, many articles and books helped inform my writing. For a full list of sources, please visit my website, www.mariakonnikova.com. Below are a few highlighted readings for each chapter. They are not intended to list every study used or every psychologist whose work helped shaped the writing, but rather to highlight some key books and researchers in each area.

Prelude

For those interested in a more detailed history of mindfulness and its impact, I would recommend Ellen Langer's classic *Mindfulness*. Langer has also published an update to her original work, *Counterclockwise: Mindful Health and the Power of Possibility*.

For an integrated discussion of the mind, its evolution, and its natural abilities, there are few better sources than Steven Pinker's *The Blank Slate* and *How the Mind Works*.

Chapter One: The Scientific Method of the Mind

For the history of Sherlock Holmes and the background of the Conan Doyle stories and Sir Arthur Conan Doyle's life, I've drawn heavily on several sources: Leslie Klinger's *The New Annotated Sherlock Holmes*; Andrew Lycett's *The Man Who Created Sherlock Holmes*; and John Lellenerg, Daniel Stashower, and Charles Foley's *Arthur Conan Doyle: A Life in Letters*. While the latter two form a compendium of information on Conan Doyle's life, the former is the single best source on the background for and various interpretations of the Holmes canon.

For a taste of early psychology, I recommend William James's classic text, *The Principles of Psychology*. For a discussion of the scientific method and its history, Thomas Kuhn's *The Structure of Scientific Revolutions*. Much of the discussion of motivation, learning, and expertise draws on the research of Angela Duckworth, Ellen Winner (author of *Gifted Children: Myths and Realities*), and K. Anders Ericsson (author of *The Road to Excellence*). The chapter also owes a debt to the work of Daniel Gilbert.

Chapter Two: The Brain Attic

One of the best existing summaries of the research on memory is Eric Kandel's *In Search of Memory*. Also excellent is Daniel Schacter's *The Seven Sins of Memory*.

John Bargh continues to be the leading authority on priming and its effects on behavior. The chapter also draws inspiration from the work of Solomon Asch and Alexander Todorov and the joint research of Norbert Schwarz and Gerald Clore. A compilation of research on the IAT is available via the lab of Mahzarin Banaji.

Chapter Three: Stocking the Brain Attic

The seminal work on the brain's default network, resting state, and intrinsic natural activity and attentional disposition was conducted by Marcus Raichle. For a discussion of attention, inattentional blindness, and how our senses can lead us astray, I recommend Christopher Chabris and Daniel Simon's *The Invisible Gorilla*. For an in-depth look at the brain's inbuilt cognitive biases, Daniel Kahneman's *Thinking, Fast and Slow*. The correctional model of observation is taken from the work of Daniel Gilbert.

Chapter Four: Exploring the Brain Attic

For an overview of the nature of creativity, imagination, and insight, I recommend the work of Mihaly Csikszentmihalyi, including his books *Creativity: Flow and the Psychology of Discovery and Invention* and *Flow: The Psychology of Optimal Experience*. The discussion of distance and its role in the creative process was influenced by the work of Yaacov Trope and Ethan Kross. The chapter as a whole owes a debt to the writings of Richard Feynman and Albert Einstein.

Chapter Five: Navigating the Brain Attic

My understanding of the disconnect between objective reality and subjective experience and interpretation was profoundly influenced by the work of Richard Nisbett and Timothy Wilson, including their groundbreaking 1977 paper, "Telling More Than We Can Know." An excellent summary of their work can be found in Wilson's book, *Strangers to Ourselves*, and a new perspective is offered by David Eagleman's *Incognito: The Secret Lives of the Brain*.

The work on split-brain patients was pioneered by Roger Sperry and Michael Gazzaniga. For more on its implications, I recommend Gazzaniga's *Who's in Charge?: Free Will and the Science of the Brain*.

For a discussion of how biases can affect our deduction, I point you once more to Daniel Kahneman's *Thinking, Fast and Slow*. Elizabeth Loftus and Katherine Ketcham's *Witness for the Defense* is an excellent starting point for learning more about the difficulty of objective perception and subsequent recall and deduction.

Chapter Six: Maintaining the Brain Attic

For a discussion of learning in the brain, I once more refer you to Daniel Schacter's work, including his book *Searching for Memory*. Charles Duhigg's *The Power of Habit* offers a detailed overview of habit formation, habit change, and why it is so easy to get stuck in old ways. For more on the emergence of overconfidence, I suggest Joseph Hallinan's *Why We Make Mistakes* and Carol Tavris's *Mistakes Were Made (But Not by Me)*. Much of the work on proneness to overconfidence and illusions of control was pioneered by Ellen Langer (see "Prelude").

Chapter Seven: The Dynamic Attic

This chapter is an overview of the entire book, and while a number of studies went into its writing, there is no specific further reading.

Chapter Eight: We're Only Human

For more on Conan Doyle, Spiritualism, and the Cottingley Fairies, I refer you once more to the sources on the author's life listed in chapter one. For those interested in the history of Spiritualism, I recommend William James's *The Will to Believe and Other Essays in Popular Philosophy*.

Jonathan Haidt's *The Righteous Mind* provides a discussion of the difficulty of challenging our own beliefs.

Postlude

Carol Dweck's work on the importance of mindset is summarized in her book *Mindset*. On a consideration of the importance of motivation, see Daniel Pink's *Drive*.

INDEX